摩登样板间 II
Modern Sample Houses II

后现代新古典
Post-Modern Neo-Classical Style

ID Book 工作室 编

华中科技大学出版社
http://www.hustp.com
中国·武汉

The Beauty of Neo-Classicism

For quite a long time, people have never halted their pursuit of art. In the art movement emerging one after another, under the guidance of common tenets and purpose, the artists affect the developing orientation of world culture with the posture of avant-guard thinking. The Neo-Classicism has profound influence till today. It is dated back to the mid of 18th century, when people were against the complicated decorative style of Rococo and aspired to go back to the age with pure art of ancient Greece and ancient Rome.

Neo-Classical style influences a lot of areas such as decorative art, architecture, painting, literature, etc. After several centuries, this style is still as dazzling and magnificent as pearl. After being in the industry of interior design for more than 10 years, I am profoundly motivated by the arousing of inspirations by this style and the everlasting mastering of beauty. The design essence of loose form but condensed spirit aroused the emotions from the bottom of people's heart. During the design process, I have always been pursuing perfect graphic composition and making use of magnificent, elegant and epic like visual effects to move people's heart. This pure attitude in the creation is integrated with the honing of life, which is finally represented in the work itself.

With the current social background, the beauty of traditional spiritual temperament is retained and there appears the visual impression of extreme novelty. What is more, Neo-Classicism is retained as some renewed beauty which plays some irreplaceable function in the art current, producing some abundant and various creative languages for modern design field.

Variations with positive significance is always comforting. We are quite satisfied with the fact that there is the trend for Neo-Classicism in the time. A good design must can add some personal creative ideas for the more profound aesthetic thinking. And that is where the quality of Neo-Classicism lies.

Liu Weijun
PINKI Interior Design & IARI Interior Design Co., Ltd.

前言

新古典主义之美

长久以来，人类对于艺术的追求一直没有停止过。层出不穷的艺术运动中，艺术家在共同的宗旨和目标指引下，以思想先锋的姿态影响着世界文化的走向。至今仍具有深刻影响的新古典主义始于18世纪中叶，反映了人们对洛可可繁复装饰风格的反对，渴望回到古希腊和古罗马时期艺术的"纯洁"。

新古典主义风格影响了装饰艺术、建筑、绘画、文学等众多领域。在经历几个世纪以后，依然如珍珠般华彩万千。在室内设计行业从业十余年来，我对新古典主义带给我灵感的催生和对恒久美的把握深有感怀。其"形散神聚"的设计精髓能够唤起人们内心深处的情感。我在设计过程中，也一直在追求完美的构图，运用大气、典雅、史诗般的视觉效果打动人的内心。这种创作上的纯真态度，融合了生活的磨砺迸发出来，并最终在作品中得到展现。

在现在的社会背景下，具有传统精神气质的美被保留下来，并添加进了新颖、精致的视觉感受，新古典主义被以一种"更新"之后的美而保存下来，在艺术洪流里发挥着不可替代的作用，为当代的设计领域带来丰富而多样的创作语言。

有正面意义的变化永远是令人欣慰的，我们很欣慰这个时代依然涌动着新古典主义的激浪。好的设计，一定要在深层次的美学思想上增添个性的创意，这便是新古典主义的可贵之处。

刘卫军
PINKI品伊创意集团&美国IARI刘卫军设计师事务所

CONTENTS

006
简约的纯美
Concise and Pure Beauty

014
乐城3栋B户型
Lecheng, Building 3, Flat Type B

024
乐城3栋C户型
Lecheng, Building 3, Flat Type C

030
阳江保利银滩·天际线
Poly Silver Beach·Skyline

036
睿智华庭A栋1单元
Wisdom Mansion, Building A, Unit 1

042
中海雍城世家高层样板房
High-Rise Model House of Zhonghai Yongcheng Shijia

048
永川御珑山JEEP样板房
Yongchuan Yulong Mountain JEEP Show Flat

054
南沙美福花园样板房
Nansha Meifu Garden Show Flat

060
岭兜吴宅
Lingdou Wu's Mansion

066
六都国际A户型样板房
Sixty City Show Flat Type A

072
南京彩云居B
Nanjing Iridescent Clouds Mansion B

076
喜年中心LOFT·A户型
Xinian LOFT·Flat Type A

080
喜年中心LOFT·B户型
Xinian LOFT·Flat Type B

086
瀚视界·福州泰禾红树林
Ocean Vision·Fuzhou Taihe Mangrove Forest

092
红橡华园示范单位
Sample Flat for Red Oak Mansion

096
融侨锦江悦府样板房
Rongqiao Jinjiang Yuefu Show Flat

102
中山海伦春天
Show Flat of Helen Spring in Zhongshan

108
以酷的姿态享受生活
To Enjoy Life with Some Cool Posture

112
宝安中洲中央公园样板房
Show Flat of Zhongzhou Central Park in Bao'an

116
保利叶之语样板房
Poly Yezhiyu Show Flat

120
宁波镇海维科样板房
Show Flat of Weike Property in Zhenhai of Ningbo City

124
爱尚摩登
Fashion Modern

130
逸静新贵主题样板间
YiJing Neo-Aristocracy Show Flat

134
镇江优山美地
Zhenjiang Youshan Meidi Property

140
静聆风吟·新东方寓所空间
Listen to the Chanting of Wind·New Oriental Residential Space

144
长滩美墅
Long Beach Nice Villa

148
中庚城某住宅
One Residence inside Zhonggengcheng Property

152
简
Simplicity

158
风尚
Fashion

目录

162 水晶之恋 / Crystal Love

166 九龙仓时代晶科名苑样板房 / Wharf Holdings Shidai Jingke Mingyuan Show Flat

170 凯德置地御金沙交楼标准样板房 / Yujinsha Show Flat •Kaide Land Holdings

174 香榭丽舍住宅公寓 / Champs Elysees Residential Apartment

180 江南人家 / Jiangnan Residence

184 光与影的舞蹈 / Dance of Light and Shadow

188 保利东语花园样板房•水韵 / Poly Dongyu Garden Show Flat•Charms of Water

194 镇江科苑华庭 / Zhenjiang Keyuan Mansion

200 微设计系列之暮光 / Evening Twilight of Micro Design Series

206 和弦悠扬 / Melodious Chords

210 白色的暖意 / White Warmth

214 黑与白的邂逅 / The Encountering of Black and White

218 赏•秋 / Appreciation•Autumn

222 听•春 / Listen•Spring

226 唤醒生活 / Wake-up Life

230 跳跃的精灵 / Dancing Angles

234 光大锦绣山河样板房 / Show Flat of Ever-bright Splendid Landscape Property

238 卡门花园 / Carmen Garden

242 纯色心情 242 / Pure Moods

246 纯色样板生活 / Pure Sample Life

250 长宁复邦住宅 / Fubang Residence in Changning District

254 名都城住宅 / Mingducheng Residence

258 清水湾住宅 / Clear Water Bay Residence

262 万科缤纷夏日样板房 / Show Flat of Vanke's Wonderful Summer Property

266 五矿•哈施塔特别墅样板房 / Wukuang•Hallstatt Villa Show Flat

270 后现代装饰主义样板间 / Show Flat of Post-Modern Decorationism

274 宜家风格样板房 / Show Flat of IKEA Style

278 摩登个性之家 / Modern and Personal Home

282 布吉私人自建别墅 / Buji Private Villa

288 眷鸟 / Juan Niao

294 苏州水岸枫情公寓 / Riverside Maple Apartment in Suzhou

300 常熟世茂三期样板间 / Show Flat of 3rd Phase of Shimao Property in Changshu

设计单位：深圳市尚邦装饰设计工程有限公司	Design Company: Shenzhen Shangbang Decorative Design and Engineering Co., Ltd.
设 计 师：潘旭强、刘均如	Designers: Pan Xuqiang, Liu Junru
项目地点：四川省成都市	Project Location: Chengdu in Sichuan Province
项目面积：150 m²	Project Area: 150 m²
摄 影 师：林力	Photographer: Lin Li

看惯了艳丽的色彩，缤纷的装饰，设计师更想营造出一个洗尽铅华后的素雅空间。在本案的设计中，空间的色调搭配以清新淡雅的颜色为主，肃静的浅褐色地面、乳白色的墙面石材、木色调的木饰面板、充满艺术气息的墙面镜、条纹相见的壁纸、格调高雅大气的格子形地毯，都为本空间增添了一丝优雅感。

在家具陈设上，造型简约的沙发极富时尚气息，造型雅致的桌面摆饰为室内空间增添了趣味感，奢华大气的水晶吊灯散发出柔和的光芒。这一切都使得室内空间充满了一种简约、高雅却低调的时尚感。这正是设计师以及业主所极力想营造的室内气息。

Being tired of gorgeous colors and splendid decorations, the designer tries to create an elegant space with all the magnificence gone. For the design of this project, the color collocation of the space is focused on fresh and delicate colors. The solemn light brown floor, milky white wall stone, wood color veneer, artistic wall mirror, wallpaper of stripes and elegant and magnificent lattice carpet all add some elegant atmosphere for the space.

For the layout of the furniture, the sofa of concise format creates some fashionable atmosphere. The desk accessories of nice shape produce some interests for the interior space. The crystal chandeliers of luxury and magnificence send out some soft lights. All these make the interior space full of concise, decent but low-profile fashionable atmosphere. And that is the interior atmosphere that the designer and the property owner want to create.

乐城3栋B户型
Lecheng, Building 3, Flat Type B

设计单位：戴维斯室内装饰设计（深圳）有限公司
设 计 师：Thomas（HK）、Marco
项目地点：广东省深圳市
项目面积：90 m²
主要材料：灰镜、清镜、黑镜钢、木饰面、木地板、扪皮、扪布、特色玻璃、水晶灯饰、壁纸、大型花卉、陶瓷锦砖、云石、地毯

Design Company: Davis Interior Decorative Design (Shenzhen) Co., Ltd.
Designers: Thomas(HK), Marco
Project Location: Shenzhen in Guangdong Province
Project Area: 90 m²
Major Materials: Grey Mirror, Clear Mirror, Black Mirror Steel, Wood Veneer, Leather, Wood Floor, Special Glass, Crystal Lighting, Wallpaper, Flowers, Ceramic Mosaic Tile, Marble, Carpet

本案的整体装饰设计为现代简约风格，以白色为基调，具有现代气息、充满活力的深蓝色作为延伸。背景墙采用简洁而明快的线条缔造出动感的造型，并利用镜钢这一特定材料将整个立面空间进行划分，点、线、面各个元素充分结合，使其达到一种高度融合的状态。再配合具有独特个性的家具，体现出现代都市人追求个性的需求，从另一个角度全新定位了家对现代人生活的不同感受。

The whole decorative style of this project is modern concise style, with white as the tone color and dynamic blue of modern atmosphere as the spreading color. The background wall applies concise and brisk lines to produce energetic format. The designer makes use of a specific materials, i.e. Mirror steel, to divide the whole space. With the complete combination of spots, lines and surfaces, some status of high integration is achieved. Accompanied with furniture of peculiar features, representing modern urban people's requirements for personalities. Thus from some other angle, modern people's different sensations for a home are oriented in a brand new way.

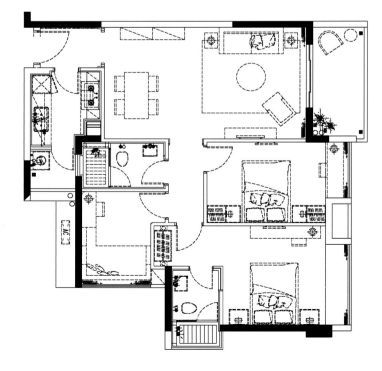

乐城3栋C户型

Lecheng, Building 3, Flat Type C

设计单位：戴维斯室内装饰设计（深圳）有限公司
设 计 师：Thomas（HK）、Marco
项目地点：广东省深圳市龙岗区
项目面积：85 m²
主要材料：拉丝钢、金镜钢、金镜、清镜、木地板、扣皮、扣布、特色玻璃、水晶灯饰、壁纸、大型花卉、陶瓷锦砖、云石、地毯

Design Company: Davis Interior Decorative Design (Shenzhen) Co., Ltd.
Designers: Thomas(HK), Marco
Project Location: Longgang District, Shenzhen in Guangdong Province
Project Area: 85 m²
Major Materials: Brushed Steel, Gold Mirror Steel, Gold Mirror, Clear Mirror, Wood Floor, Leather, Special Glass, Crystal Lighting, Wallpaper, Flowers, Ceramic Mosaic Tile, Marble, Carpet

后现代新古典

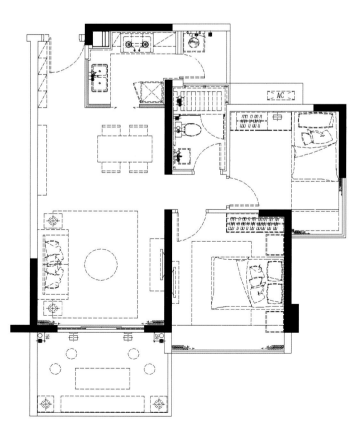

本案的整体装饰格调为现代简约风格，以浅色调为主，搭配少量的黑色以及金色、绿色，混搭出清新且时尚的格调。

在材料的运用上，设计师将大量的现代材质应用进来，镜面、扣皮、瓷砖、墙纸等材质的运用，提升了整个居室空间的品味。水晶灯饰、不锈钢拉丝家具、各式花卉的装点又为居室空间增添了一些温柔的气息。体现出现代都市人对个性、时尚品味生活的追求。

The whole decorative tone of this project is modern concise style, majored on light color tone, accompanied with some black, gold and green colors, creating some fresh and fashionable tones.

As for the application of materials, the designer makes use of a lot of modern materials, such as mirror surface, leather, tile and wallpaper which uplift the taste of the residential space. The decoration of crystal lights, stainless steel brushed furniture and various flowers add some gentle atmosphere for the space. The design displays modern urban people's pursuit for personality and fashionable life of high taste.

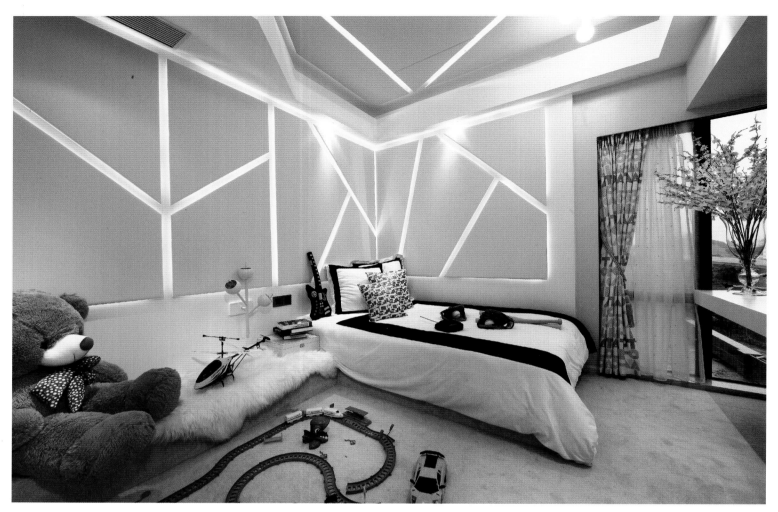

阳江保利银滩·天际线
Poly Silver Beach·Skyline

设计公司：广州道胜装饰设计有限公司
设 计 师：何永明
项目面积：162 m²
主要材料：白色微晶石、白色墙纸、白色烤漆木饰面、雅士白大理石、米黄色地毯、金黄色软包
摄 影 师：彭宇宪

Design Company: Guangzhou Daosheng Decorative Design Co., Ltd.
Designer: He Yongming
Project Area: 162 m²
Major Materials: White Microlite, White Wallpaper, White Wood Veneer of Baking Finish, Jazz White Marble, Beige Carpet, Beige Soft Roll
Photographer: Peng Xianyu

本项目所处的位置毗邻一个高尔夫球场，所以我们对本项目的定位就是以营造休闲度假感为主的高端别墅，从而提高该项目的档次及品位。

该设计主要是运用简约的弧线造型，弧线贯穿于空间的每一个角落。明快的色彩体现出愉悦的生活氛围，柔和的米白色调中加上适量的橙色做点缀，不仅活跃了空间气氛，也使得整个空间呈现出一种休闲度假的感觉。

另外，空间引入了大量的高尔夫元素，与整个户外环境相辅相成。简洁的家具在空间中也能形成丰富的层次关系，延续了在硬装上的设计元素，柔和的弧线也体现在每件家具上；搭配地脉图案的地毯，增强了视觉的效果；挂画选用自然的风景油画，营造出休闲的度假氛围，高尔夫元素的立体画，更突显出本案的主题。

本套样板房的设计过程更像是一种多元化的思考方式，设计师用理性、细腻的设计语言，解构和整合空间主题，将大胆的创意与现代生活相结合，体现出主人的高品位和休闲放松的生活氛围。

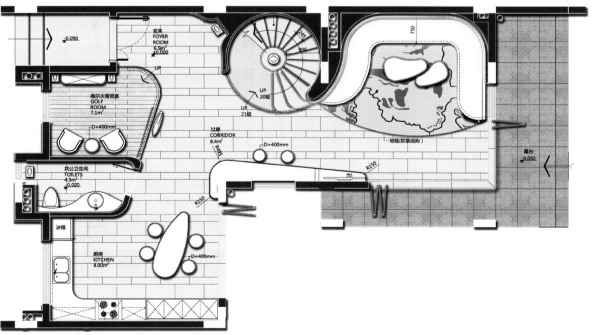

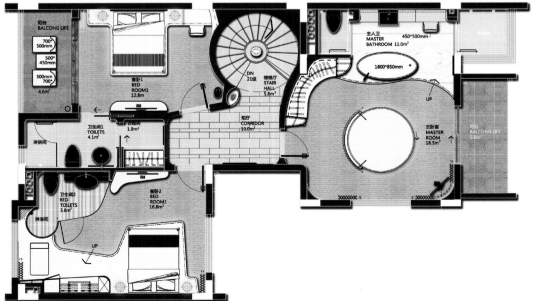

This project is located in an area neighboring a golf course. We orientated this project to be a high-end villa for creating leisure holiday feel, thus, promoting the standard and taste of this project.

This design mainly makes use of arc pattern which runs through every corner of the space. The bright colors display some pleasing living atmosphere. The soft beige white color tone is added with some proper orange colors, which not only activates the space atmosphere, but also makes the whole space send out some leisure holiday sensations.

Other than that, the space makes use of quite a lot of golf elements, which are in accordance with

the whole outdoor environment. The concise furniture can produce some abundant layers inside the space, continuing the design elements in the hard decorations. The soft arc lines are also displayed in every piece of furniture. Accompanied with the carpet of graphics, the visual effects are strengthened. The painting hung on the wall is a natural landscape painting, creating some leisurely holiday atmosphere. The three-dimensional painting of golf elements also highlights the theme of this project.

The design process of this show flat is like some diversified thinking modes. The designer makes use of the reasonable and exquisite design language to deconstruct and integrate the theme of the space, combining bold creative ideas with modern life, representing the high taste of the property owner and the leisure and relaxing life atmosphere.

睿智华庭A栋1单元
Wisdom Mansion, Building A, Unit 1

设计单位：戴维斯室内装饰设计（深圳）有限公司
设 计 师：Thomas（HK）、Marco
项目地点：广东省深圳市
项目面积：90 m²
主要材料：灰镜、清镜、黑镜钢、木饰面、木地板、扪皮、扪布、特色玻璃、水晶灯饰、壁纸、大型花卉、陶瓷锦砖、云石、地毯

Design Company: Davis Interior Decorative Design (Shenzhen) Co., Ltd.
Designers: Thomas(HK), Marco
Project Location: Shenzhen in Guangdong Province
Project Area: 90 m²
Major Materials: Grey Mirror, Clear Mirror, Black Mirror Steel, Wood Veneer, Leather, Wood Floor, Special Glass, Crystal Lighting, Wallpaper, Flowers, Ceramic Mosaic Tile, Marble, Carpet

本案设计以现代温馨为主题,设计师将白色作为主体色调,背景墙采用简洁的立体造型。并利用少量金镜的点缀来衬托时尚华丽的家具,体现出都市人对时尚生活的追求。整个设计包含了三种现代气息。

现代简欧:以白色的简欧风格的线条,勾勒出富有欧式氛围的空间感。现代与欧式风格混搭的家具,体现出客户对品味生活的追求。

现代奢华:奢华的灯具在造型上却非常简洁,香槟金优雅的点缀其间,银镜的巧妙运用营造出丰富的空间层次,玻璃与灯光的奇妙结合营造出梦幻般的空间氛围。

现代贵族:传统风格与奢华的格调诠释出别样的贵族风情,再点缀上古色古香的茶色,传统意蕴中华丽尽显。

The design is focused on being modern and warm. The designer has white as the tone color, while the background wall applies concise three-dimensional formats. The designer applies some gold mirror to set off the fashionable and magnificent furniture, representing urban people's longing for fashionable life. The whole design includes three modern styles.

Modern concise European style. With white lining of concise European style, the space feel of European atmosphere is created. The furniture combining modern and European styles displays the customers' pursuit of life quality.

Modern luxury style. The luxurious lights are quite concise in the format. Dotted with champagne gold ornaments, the ingenious application of silver mirror creates abundant space layers. The magical combination of glass and lights produces dreamy space atmosphere.

Modern aristocratic style. Traditional style and luxurious tone display some peculiar aristocratic charms. Accompanied with the antique tea color, there is splendor in the traditional connotations.

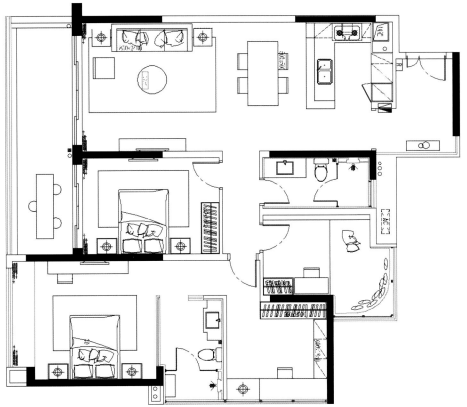

中海雍城世家高层样板房
High-Rise Model House of Zhonghai Yongcheng Shijia

设计单位：上海大研室内设计工程有限公司	Design Company: Shanghai Dayan Interior Design and Engineering Co., Ltd.
设 计 师：欧阳辉	Designer: Ouyang Hui
参与设计：姚君、顾莉莉	Associate Designers: Yao Jun, Gu Lili
项目地点：浙江省宁波市	Project Location: Ningbo in Zhejiang Province
项目面积：173 m²	Project Area: 173 m²
主要材料：乳胶漆、橡木染灰、大理石、镀钛不锈钢、超白玻璃	Major Materials: Emulsion Paint, Oak Ash, Marble, Titanized Stainless Steel, White Glass

宁波是座历史文化古城，同时也是旅游城。本项目位于宁波市鄞州区，是中海集团精心打造的雍城世家项目中的高层样板房。

作为城市居住空间的典范，雍城世家样板房的设计也要针对特定目标群体的需求而定。设计手法以现代都市生活的多变性和多功能性来表现。进门可以看到一件具有东方风情的可移动屏风，与客厅中的电视背景墙相呼应，电视墙面采用了肌理丰富的石材，自然的朽木与电视墙相结合。大面积的烤漆材料的投影幕使线和面结合起来。

餐厅中超长的实木餐桌给人以大气、奢华的印象。在不宽的走道中设计师用具有半通透质感的石材作为地面主材，让原本狭长的走道在光影的衬托下显得丰富起来。卫浴间中定制的镀钛不锈钢台盆显得现代而时尚，卫浴柜使用了镜面材质，使得空间的尺度得到了扩展，另外卫浴间的墙面还使用了钢化玻璃，使得空间更显通透。

多功能儿童房拥有供孩子学习、游戏以及睡眠等多项功能，设计师巧妙地在墙面、地面、柜体及踢脚线等处运用线形的 LED 灯，满足儿童向往多彩生活的心理需求

Ningbo is not only a historical and cultural ancient town, but also a tourist city. This project is located in Yinzhou District of Ningbo City and is a high-rise model house that Zhonghai Group deliberately creates for Yongcheng Shijia Project.

As a model for urban living space, the design should be oriented according to the requirements of specific target groups. The design approach is represented with the varying characteristics and multi-functional characteristics of modern urban life. Upon entering the door, you can find a movable screen with oriental charms corresponding with the TV background wall of the living room. The TV background wall applies stones with rich texture and the natural deadwood is combined with the TV background wall. The projection screen of large area baking finish combines lines and surfaces.

The long solid wood dining table inside the dining hall leaves people with magnificent and luxurious impression. In the not too broad corridor, the designer applies stones with semi-transparent texture as the major material for the floor, and the original long corridor is made resplendent set off by light and shadow. The custom-made titanized stainless steel basin appears modern and fashionable. The cupboard inside the washroom applies mirror surface materials to expand the scale of the space. Other than that, the wall of the washroom makes use of toughened glass to make the space appear transparent.

Multi-functional children's room has various functions for them to learn, play games and sleep here. The designer ingeniously applies linear LED lights on wall, ground, cabinet and leg wire, etc. to meet children's psychological requirements for colorful life.

永川御珑山 JEEP 样板房
Yongchuan Yulong Mountain JEEP Show Flat

设计单位：戴勇室内设计师事务所
项目面积：70 m²
项目地点：重庆
主要材料：毛石、云石、皮革

Design Company: Eric Tai Interior Design Firm
Project Area: 70 m²
Project Location: Chongqing
Major Materials: Quarry Stone, Marble, Leather

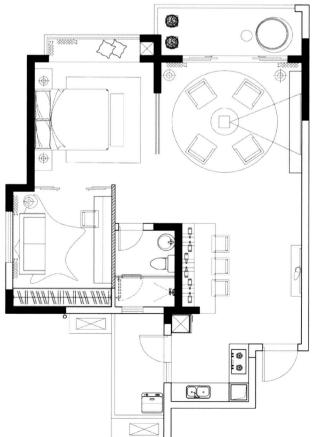

开着吉普驰骋在一望无际的高原上，追赶着夕阳下的牛群。旷野的风从半开的车窗外呼呼刮过，天高地阔，自由在呼唤着不羁的心奔向新的旅程……

本案以一个爱好吉普车的时尚男士为设计蓝本，营造出不一般的空间感受。大面积的毛石墙面体现出一种原始的粗犷质感，深沉稳重的色调和灯光效果演绎出男主人的成熟魅力。客厅中用越野车车轮打造的茶几和时尚的椅子，更是给人超酷的感受。

一顶咖啡色的牛仔帽搁在床上，书架上摆放着巴顿将军的黑白照片和军用水壶，透露出主人对旅行和军事的喜爱。深色皮革的床头背景墙，大花纹的啡色云石地面，都是对沉稳、专注和坚毅的男性魅力的间接描述。

在这个 70 m² 的公寓里，设计师不仅为我们展示了独具特色的"JEEP"风格，更为我们展示出一位时尚男主人的独特个性和品味。勇敢洒脱的人会选择自己热爱的生活方式，活出属于自己的风格。习惯了城市里循规蹈矩、按部就班的生活，何不偶尔放松一下，给自己一个自由随性的机会。

趁着户外秋高气爽的好天气，开着吉普向野外出发吧……

He is driving a jeep on a vast plateau, chasing oxen in the dusk. The wild wind blows by from half-open car windows, and his unbounded mind is set free to sail on his new journey, under the high sky and on the broad land…

The development takes a Jeep-fancier as design blueprint to create special feeling in the space. The rubble wall presents a sense of original roughness, and the deep, stable color tone and light effect reveal the mature charm of the host. In the parlor, the end table decorated with off-road wheel and fashionable chairs leaves people with super-cool impressions.

On the bed lays a coffee color cowboy hat, and on the shelf are General Barton's black-and-white photograph and a canteen, implying the host's hobbies of traveling and military stuff. Both the bed background with brunet leather and the floor with coffee-strip marble indicate masculine charm of calmness, dedication, and persistence.

In the 70 m^2 of the apartment, the designer displays to us not only the particular Jeep style but also the unique individuality and taste of the host. Brave and free man will choose to embrace the life style of his own. Now that you have accustomed to the rules of city life, why not set yourself free and give yourself a chance to live your own life for one moment?

Don't miss the time of cool winter, and drive a jeep to the wild!

南沙美福花园样板房
Nansha Meifu Garden Show Flat

设计单位：斯考利王装饰设计有限公司
设 计 师：Roger B
项目地点：广东省广州市

Design Company: Sikao Liwang Decorative Design Co., Ltd.
Designer: Roger B
Project Location: Guangzhou in Guangdong Province

在本案的设计中，设计师将简约风格与中国居住文化的设计理念相结合，特别清楚地强调出简约的内涵。在设计中，充分地考虑到了空间的功能性以及对储物空间的需求，将各种储物空间巧妙地藏起来，却又保持了表面的美观，迎合了中国家庭的居住需求。

因为在设计之前与客户进行了充分的沟通，因此对客户分析非常的详尽到位，使得客户需求的重点得到充分考虑，并很好地实现了用户的想法。

在灯光设计上，光线让视线得到最大化的延伸，丰富了空间的层次。在功能设计上勾勒出了一个简单而颇具个性的现代化的生活空间，是一个满含设计师热情与智慧的设计作品。

For the design of this project, the designer combines concise style with the design concepts of Chinese residential culture, clearly emphasizing the connotations of conciseness. In the design, the designer takes full account of the functionality of space and requirements towards storage. All the storage spaces are ingeniously concealed, maintaining the surface appearance while meeting with residential demands of Chinese families.

The designer carries out sufficient negotiation with the customers, having detailed analysis towards the customers and enforcing the preferences of the customers.

For the lights design, the lights maximizes the lines of sight, while enriching the layers of the space. In functional design, the designer delineates a simple but peculiar modern residential space. This is a piece of design work full of the designer's enthusiasm and wisdom.

岭兜吴宅
Lingdou Wu's Mansion

设 计 师：张坚
设计团队：刘可华、邱纪蒙、吴小云
项目地点：福建省厦门市
项目面积：130 m²
主要材料：霸王花大理石、银雕大理石、灰橡面板、木地板、米色玻化砖
摄 影 师：许晓东

Designer: Zhang Jian
Design Team: Liu Kehua, Qiu Jiying, Wu Xiaoyun
Project Location: Xiamen in Fujian Province
Project Area: 130 m²
Major Materials: Marble, Grey Oak Veneer, Wood Flooring, Beige Vitrified Tile
Photographer: Xu Xiaodong

后现代新古典

本案例的业主为一对年轻夫妇，喜欢追求时尚、简练的生活氛围。在风格与色调的把握上，设计师与业主进行了详细的沟通，并取得了一致的意见。本案的设计风格以现代简约为主，略带点东方格调。在色彩上，设计师用灰色来定义这个空间的基调，配上一些米灰色的材质作为过渡，使整个空间充满个性却又不乏温馨感。

设计师力求让每个进入房子的人第一眼就能感受到空间的宽阔感，因此在书房与公共空间的隔断上使用了玻璃与黑钛拉丝不锈钢的组合，使得客厅、餐厅、书房融为一体。卧室以浅色调为主，地面主材为木地板，大片落地窗的采光使卧室充满休闲的气氛，衣柜与电视背景墙的结合也让卧室空间显得更加简洁、大方。

The property owner of this project is a young couple who likes fashionable and simple life atmosphere. For the aspect of style and tone, the designer carries out detailed communication with the property owner and achieves consensus. The design style of this project focuses on modern conciseness, with a little bit oriental charm. As for color, the designer chose grey to define the tone of the space. Accompanied with some gray materials as the transition, the whole space possesses individuality and warmth.

The designer tries to make everyone perceive the spaciousness of the space upon entering the space. Thus the partition between the study and the public space applies the combination of glass and black titanium stainless steel. Hence the living room, dining hall and the study are integrated as a whole. The bedroom focused on light color tone. The major materials for the floor are wood flooring. The natural lighting of the large French window fills the bedroom with leisurely atmosphere. And the combination of wardrobe and TV background wall makes the bedroom appear concise and magnificent.

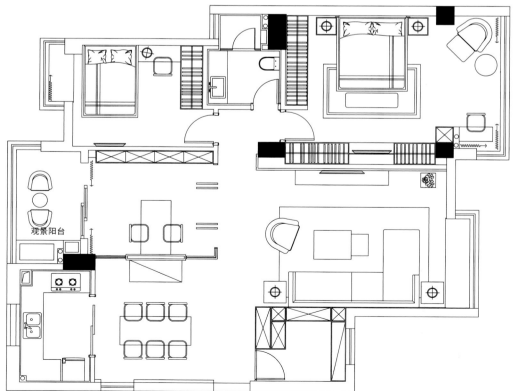

设计单位：鸿扬集团 · 陈志斌设计事务所
主设计师：陈志斌
参与设计：谢琦
项目地点：湖南省长沙市
项目面积：121 m²
主要材料：闪电米黄、银箔、皮质软包、壁纸
摄影师：管盼星

Design Company: Chen Zhibin Design Studio of Hirun Group
Chief Designer: Chen Zhibin
Associate Designer: Xie Qi
Project Location: Changsha in Hunan Province
Project Area: 121 m²
Major Materials: Lightning Beige Marble, Silver Foil, Leather Soft Roll, Wallpaper
Photographer: Guan Panxing

后现代新古典

本案为后现代奢华风格。原始空间布局合理，面积适中。设计师用现代的设计手法充分演绎出了奢华空间的高贵品质。

客厅墙面通过米黄石材、银镜的质感以及色调的对比，体现出不同材质带给人的不同感受。顶部的藻井式天花贴上了银箔壁纸，色调上与整体空间统一起来。沙发背景墙上用乳白色的软包来装饰，再配上不落俗套的挂饰和家具，把现代感和高贵感都呈现出来。

主卧室与卫浴相连，墙面上贴满带有印花纹样的壁纸，窗帘的花色也充满了浓浓的怀旧风情，床头背景墙用暗红色、不规则的硬包拼贴在一起，设计师还特意在这个空间中设置了一个半封闭的小阳台作为休闲空间，放置一张小桌、一把椅子，可以在此品茗、阅读，为生活增添了一番情调。

This project is post-modern luxurious style. The original space has appropriate layout and size. With modern design approaches, the designer fully interprets the noble temperament of luxurious space.

Through beige stone, silver mirror with texture and the contrast of color tones, the living room's wall displays different impressions that different materials leave people with. The caisson ceiling on the top is paved with silver wallpaper which unifies with the whole space in color tone. The sofa's background wall is decorated with milk white soft roll. Accompanied with uncommon accessories and furniture, both modern feel and aristocratic feel are displayed.

The master bedroom is connected with the bathroom. The wall is all posted with printing pattern wallpaper. The pattern of curtains also produces intensive nostalgia charms. The bedside background wall puts together dark red and irregular hard rolls. The designer specially sets a semi-enclosed tiny balcony inside this space as the leisure space. There are also a table and a desk. One can read books and enjoy tea here. This adds some emotional appeals for life.

摩登样板间 II

南京彩云居 B
Nanjing Iridescent Clouds Mansion B

设计单位：大匀国际空间设计
设 计 师：林宪政、王华
软装设计：太舍馆贸易有限公司
项目地点：江苏省南京市
项目面积：90 m²
主要材料：橡木地板、人造橡木饰面板、细花白石材、米白乳胶漆

Design Company: Symmetry Space Design
Designers: Lin Xianzheng, Wang Hua
Soft Decoration Design: MOGADECO Co., Ltd.
Project Location: Nanjing in Jiangsu Province
Project Area: 90 m²
Major Materials: Oak Flooring, Manmade Oak Veneer, White Stone, Creamy White Emulsion Paint

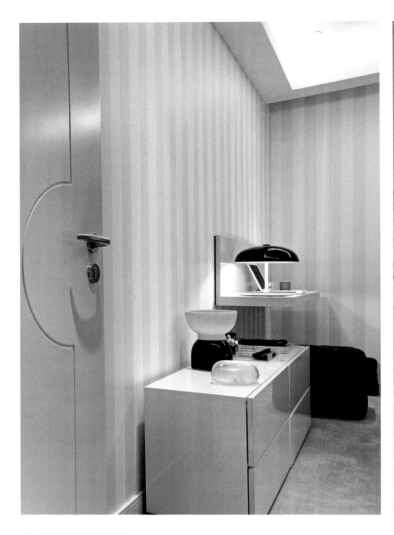

比"宜家"更温馨

这是一个中小面积的样板间,如何在有限的空间内打造出舒适、宜家且大气的室内空间氛围是设计师要思索的问题。

设计师采用了现代风格的设计格调,整体色调以浅咖色、乳白色为主。沙发背景墙采用了具有麻面质感的壁布,墙面上交错放置的风景画给居室增添了一些自然的气息。沙发的样式简约而时尚,地面上铺设了咖色的毛绒地毯。形态纤小、可爱的白色茶几上放置了必备的白色茶壶、茶盅。花瓶中插着一支白色鸢尾,其优雅的姿态给人以清新的感觉。

餐厅的设计延续了客厅简练、舒畅的设计风格,富有艺术气息而形态简洁的吊灯散发出光晕柔和的光芒,白色的橡木地板在空间内延伸,餐柜的造型简洁而大气。

Warmer than "IKEA"

This is a small and medium-sized sample house. The designer needs to address the issue of creating comfortable, cozy and magnificent space atmosphere with the limited space.

The designer applies modern-style design tone. The whole color tone focuses on light coffee color and milk white. The sofa's background wall makes use of wall cloth with pitted surface. The landscape paintings arranged in some staggered way create some natural atmosphere. The pattern of sofa is concise and fashionable. And the floor is paved with coffee color pile floor covering. The necessary white teapot and tea cups are placed on the tiny and cute white tea table. There is a white fleur-de-lis in the vase which has elegant posture giving people fresh impressions.

The design of the dining hall inherits the concise and comforting design style of the living room. The droplight full of artistic breath which has concise format sends our rays of light with soft halos. White oak flooring extends in the space. And the format of cabinet is simple but magnificent.

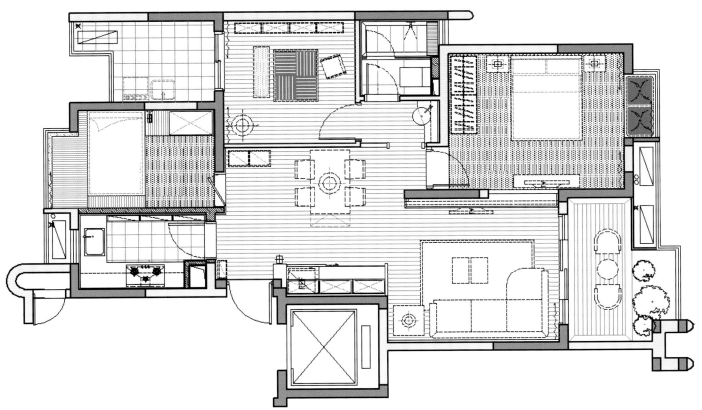

喜年中心 LOFT·A户型
Xinian LOFT·Flat Type A

设计单位：深圳市昊泽空间设计有限公司	Design Company: Shenzhen Haoze Space Design Co., Ltd.
设 计 师：韩松	Designer: Han Song
项目地点：江苏省无锡市	Project Location: Wuxi in Jiangsu Province
项目面积：50 m²	Project Area: 50 m²
主要材料：木地板、石材、银镜	Major Materials: Wood Floor, Stone, Silver Mirror
摄　　影：江河摄影	Photographer: Jianghe Photography

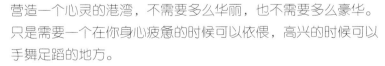

营造一个心灵的港湾，不需要多么华丽，也不需要多么豪华。只是需要一个在你身心疲惫的时候可以依偎，高兴的时候可以手舞足蹈的地方。

本案中，设计师就是要营造这样一个属于灵魂的归属地。在这个面积不大的LOFE户型的设计中，设计师在现代简约风格的基础上融入了一些装饰主义的特征，这样就给过于坚硬的外表增添了些许温情的氛围。简约的家具搭配饶有趣味的配饰，再加上稳重而又活泼的色彩，使得这个复式小居室充满着浓浓的人情味，使之真正成为一处灵魂归属地。

It does not need to be much too magnificent or luxurious by creating a harbor for the hearts. It only needs to be a place where you can lean on where you are exhausted, where you can dance to heart's contents when you are happy.

For this project, the designer does intend to create a place for the souls. For the design of this loft which is not so big in size, the designer wants to integrate some features of decoration based on modern and concise style. Thus the appearance with much too hard outlook is added with some warm atmosphere. The concise furniture is accompanied with some interesting decorative ornaments. Accompanied with sedate and dynamic colors, this duplex little residence is filled with intensive human kindness, making itself a real harbor for souls.

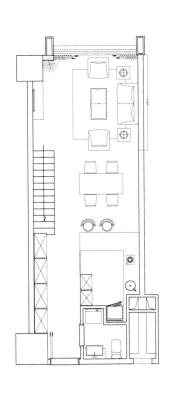
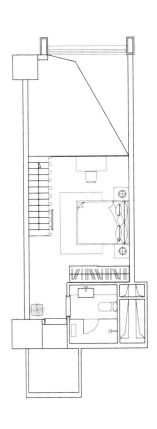

喜年中心 LOFT·B户型
Xinian LOFT·Flat Type B

设计单位：深圳市昊泽空间设计有限公司	Design Company: Shenzhen Haoze Space Design Co., Ltd.
设 计 师：韩松	Designer: Han Song
项目地点：江苏省无锡市	Project Location: Wuxi in Jiangsu Province
项目面积：50 m²	Project Area: 50 m²
主要材料：木地板、石材、银镜	Major Materials: Wood Floor, Stone, Silver Mirror
摄　　影：江河摄影	Photographer: Jianghe Photography

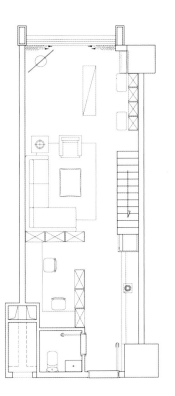
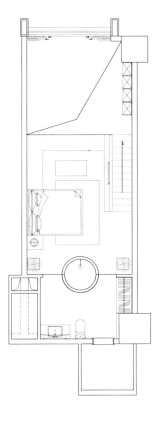

从本案可以看出居室主人浓浓的艺术情结，整面墙上挂满了富有想象力的印象画，博古架上也摆放了许多艺术装饰品，这一切都在传递出独属于这个空间的性格气质，由此可以想象居室主人的爱好以及情趣。

设计师在这个LOFT的设计中，从业主的需求出发，运用大量的艺术装置品、极具个性的装饰材料来营造浓厚的艺术氛围。在这个空间中，不仅有家的温馨感，也有极具个人情怀的氛围。在这样的居室中生活，一定会点燃生活的激情，使人们更加的热爱生活。

From this project, we can find the intensive artistic complex of the property owner. The whole wall is hung with paintings full of imaginations. There are also some decorative objects on the shelf. All these convey some characteristics and temperament exclusive to the space. From that we can get a clue of the likes and interests of the property owner.

For the design of this LOFT, the designer starts from the requirements of the property owner and makes use of quite a lot of artistic objects to create intensive artistic atmosphere. Inside this space, there are not only warm home feelings, but also atmosphere of individual feelings. Through living in this residence, people's enthusiasm for life would surely be lit and people would love life even more.

瀚视界·福州泰禾红树林
Ocean Vision · Fuzhou Taihe Mangrove Forest

设计单位：中国无印良品空间设计事务所
设计师：陈绍良
项目面积：115 m²
摄影师：施凯

Design Company: Wuyin Liangpin Space Design Firm in China
Designer: Chen Shaoliang
Project Area: 115 m²
Photographer: Shi Kai

后现代新古典

有一种空间叫做遐想，有一种情怀叫做释放。

在本案中，设计师赋予了这个空间一种精神的升华。设计师在有限的布局中采用了多个块面，用不同的设计手法，使用色块与玻璃材质等来划分空间，使得整体空间在视觉上得到了有效的扩展和延伸。通透式的设计手法贯穿于不同的单元空间中，多方位地体现出一种融洽感，这样的设计也让空间多了几分宁静与和谐。

在灯光的照射下，艺术墙折射出了亦真亦幻的梦幻魅力，置身其中的人们仿佛是在另一个奇幻世界中遨游，浪漫的气息弥漫在静谧的夜空下。

There is some kind of space which is called imagination. There is some kind of emotion called release.

In this case, the designer endows this space with some spiritual sublimation. The designer makes use of various chunks and surfaces in the limited layout. With different design approaches, the designer applies color chunks and glass materials to divide the space. Thus the whole space gets effective expansion and extension visually. Transparent design approaches run through different space units, displaying some harmony from multi aspects. This design also creates some tranquility and harmony for the space as well.

Illuminated by the lights, the art wall sends out some surreal charms. It is like that people are travelling in some other wonderland. Romantic atmosphere is pervasive under the quiet starry night.

红橡华园示范单位
Sample Flat for Red Oak Mansion

设计单位：鸿扬集团·陈志斌设计事务所
主设计师：陈志斌
参与设计：谢琦、彭辉
项目地点：湖南省长沙市
项目面积：110 m²
主要材料：爵士白石材、灰镜、壁纸、PU软包、多层实木地板
摄影师：管盼星

Design Company: Chen Zhibin Design Studio of Hirun Group
Chief Designer: Chen Zhibin
Associate Designers: Xie Qi, Peng Hui
Project Location: Changsha in Hunan Province
Project Area: 110 m²
Major Materials: Jazz White Stone, Grey Mirror, Wallpaper, PU Soft Roll, Multilayer Solid Wood Floor
Photographer: Guan Panxing

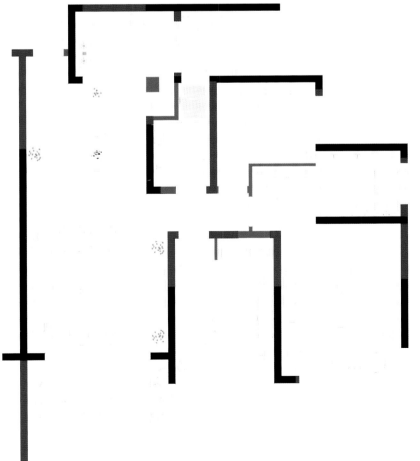

后现代新古典

在本案中，公共区域的设计以淡雅的白色为主基调，沙发背景墙上有条形的灰镜，地面上配以黑、白色块相间的皮质地毯，白色的沙发镶有黑边，电视墙上块状的爵士白石材拼接在一起，完美地勾勒出了现代人心中理想的家。

主卧室以恬静清雅的灰色为主基调，墙面上大面积铺设灰色的壁纸，床头背景墙用了咖色的条形软包，地面使用具有古朴色调的多层实木地板，烘托出家的优雅气质。次卧室的设计同样出彩，床头墙面运用乳白色的长条软包，渲染出时尚、华美的空间情调，对面墙上以黑、白条纹图案的壁纸为主，营造出淡雅、质朴、理性的氛围，同一个空间通过不同材料的特性体现出了多种情感。同时，多种软装陈设的点缀，传达出居室主人对生活的热情。

For this project, elegant white is the tone color for the design of public space. There is bar-type grey mirror on the sofa's background wall. The ground is decorated with leather carpet with black and white chunks. White sofa has black lining. The chunk jazz white stones on the TV background wall are pieced together. The ideal home for modern people is delineated in some perfect way.

The master bedroom has quiet and elegant grey color as the tone color. A large area of the wall is decorated with grey wallpaper. The bedside background wall applies brown color bar-type soft roll. The ground is decorated with multi-layered solid wood floor with primitive simplicity. All these

add shading around home with elegant temperament. The design of the secondary bedroom is also excellent, with long strip milk white soft roll on the bedside wall producing fashionable and magnificent space tones. The wallpaper on the opposite wall is mainly black and white bar-type graphics, creating elegant, primitive and rational atmosphere. Through characteristics of different materials, the same space displays multiple sentiments. At the same time, through the ornaments of various soft decorative furnishings, property owner's enthusiasm for life is displayed totally.

融侨锦江悦府样板房
Rongqiao Jinjiang Yuefu Show Flat

设计单位：林开新室内设计有限公司
设 计 师：林开新
项目地点：福建省福州市
项目面积：120 m²
主要材料：大理石、护墙板、烤漆玻璃、镜面
摄 影 师：吴永长

Design Company: Lin Kaixin Interior Design Co., Ltd.
Designer: Lin Kaixin
Project Location: Fuzhou in Fujian Province
Project Area: 120 m²
Major Materials: Marble, Wainscot Board, Baking Finish Glass, Mirror Surface
Photographer: Wu Yongchang

后现代新古典

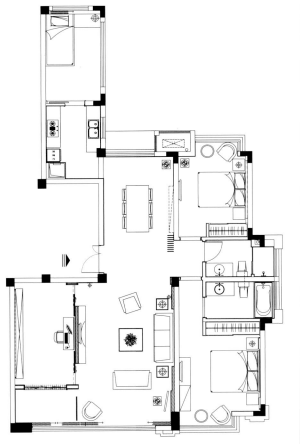

本案为福州融侨锦江新地块悦府的样板房，面积为 120 m²，四房两厅的格局。在本案的设计中，设计师结合业主的兴趣爱好、年龄、社会身份等因素，将本案打造成极具现代感，但是又富有新古典装饰情节的时尚型居所。

客厅中，造型简约的浅黄色沙发、枣红色的茶几、具有优美纹理的大理石材墙面，都使得这个空间具有超凡脱俗的雅致感。虽然没有繁复的装饰，简简单单的巧妙搭配也同样可以打造出美轮美奂的空间情境。

在整体的空间布局上，设计师将各个空间进行了很巧妙的融合，使之通透感极强，这样就使得空间的封闭性不是那么强烈，视野也随之开阔起来。

This is a show flat with four bedrooms and two living rooms. For the design of this project, the designer combines elements such as interests, age and social status of the property owner to build this project into a fashionable residence with modern feel and full of neo-classical decorative elements.

Inside the living room, the concise light yellow sofa, purplish red tea table and the marble wall of elegant grains create some outstanding elegance for the space. Although there are no complicated decorations, simple and ingenious collocations can also create a dazzling space.

For the whole space layout, the designer carries out skillful integration for each space to make it full of transparency. Thus the closure level of the space is not that intensive, while the views are also broadened.

中山海伦春天
Show Flat of Helen Spring in Zhongshan

设计单位：朗昇空间设计	Design Company: Lonson Space Design
设 计 师：袁静、钟福建	Designers: Yuan Jing, Zhong Fujian
项目面积：68 m²	Project Area: 68 m²
主要材料：大理石、玻璃、地砖、镜钢	Materials: Marble, Glass, Floor Tile, Mirror Steel

后现代新古典

小空间里的现代奢华之美

本案面积仅为 68 m²,但却有双阳台、双卧、一厨、一卫等空间,各种功能空间一应俱全。因为面积小,功能空间又多,这就需要设计师扬长避短,进行合理的规划布局,以满足主人的需要。本案采用现代风格的设计手法,注重物料品质,做工精细。设计师利用丰富的软装配饰,来表现时尚小空间中的现代奢华之美。

入户后,人们就能看到餐厅空间,墙面上大面积地使用了黑色镜钢,与精美的水晶吊灯、木纹方桌、布艺座椅、装饰画等交织在一起,给人带来一份时尚、优雅的气息。右侧的墙面位置,安放了装饰柜与黑色鞋柜,增加了空间的利用率。

客厅中,简约风格的白色沙发、黑色烤漆家具与精美的装饰品完美地搭配在一起,干净利落、时尚而现代。沙发背景墙上用大面积的木纹来装饰,鱼型饰品"跃然"其上,银白色的小鱼反射着灯光,犹如午后波光粼粼的水面上一群小鱼在欢快追逐。这使得整个空间更加活力四射。

主卧室中,设计师以黑、白两色与大面积的浅灰色相搭配,使得空间的层次丰富起来。卧室背景墙采用了灰色的方格软包,再搭配一款长方形的装饰画,时尚而别致。衣柜则使用玻璃和镜钢打造,茶色半透明的反光玻璃,延伸了视觉。

The Beauty of Modern Luxury of Small Space
The area of this residence is only 68 m², but it has multiple spaces, such as two balconies, two bedrooms, one kitchen and one bathroom. Due to the small space and the various functional spaces, this requires the designer to promote good points and avoid shortcomings, thus carrying out appropriate planning layout and meeting with the host's requirements. This project applies the design approach of modern style, emphasizes on the quality of materials and meticulous craftsmanship. The designer makes use of rich soft decoration accessories to display the beauty of modern luxury of fashionable small space.

Upon entering the space, people could perceive the dining space with large area of black mirror steel on the wall which interweaves with elegant crystal chandeliers, wood grain table, cloth chair and decorative paintings, creating some fashionable and elegant atmosphere for people. On the right side of the wall there is a decorative case and black shoe cabinet which increase the use ratio of the space.

In the living room, the white sofa of concise style integrates perfectly with furniture

of black baking finish and exquisite decorative objects, appearing clear and tidy, fashionable and modern. The sofa's background wall uses large area wood grain as the decoration, with fish-shaped ornaments on it. Silver white little fish reflects the lights, which is like a flock of fish swimming joyously on the clear water after noon. This makes the whole space much more vigorous.

The master bedroom applies the combination of black, white and large area of light grey color, enriching the layers of space. The background wall of the bedroom applies grey grid soft roll, accompanied with rectangular decorative painting, fashionable and spectacular. The wardrobe is made of glass and mirror glass. The tawny semitransparent reflective glass enlarges the vision.

以酷的姿态享受生活
To Enjoy Life with Some Cool Posture

设计单位：福州宽北装饰设计有限公司	Design Company: Fuzhou Kuanbei Decorative Design Co., Ltd.
设计师：施传峰	Designer: Shi Chuanfeng
项目面积：75 m²	Project Area: 75 m²
项目地点：福建省福州市	Project Location: Fuzhou in Fujian Province
主要材料：地板、壁纸	Major Materials: Flooring, Wallpaper
撰　　文：江雍箫	Composer: Jiang Yongxiang
摄 影 师：施凯	Photographer: Shi Kai

设计之初,主人便给这个新家烙上了鲜明的材质标签,即"不用一片瓷砖"。当我们走进这个空间时,"酷"这个字眼是对目之所及区域的最佳注解。以较少的色彩,加之金属与玻璃塑造出的冷峻,营造出居室主人的男人味。

本案中,客厅与厨房呈开放式布局,吧台成为衔接这两个功能区域的媒介。吧台的肌理与地面、电视背景墙统一起来,斑驳复古的模样略带着工业化的痕迹。这种沧桑的气质透露出一种难以抗拒的神秘感,于是这些区域自然而然地吸引着每个到此的宾客,并为之心动。吧台的椅子以现代简约的面貌营造出空间的氛围,白色金属的质感适时地缓和了厨房区域的硬气,几何线条也将装饰意味铺陈其间。我们很难将这里归类到任何一种特定风格中,这或许便是设计师的独具匠心之处吧。

毗邻厨房的客厅以黑白格调作为主旋律,黑色皮质沙发的体量感在这种氛围的家居空间中显得十分到位,有种扎实的、落地生根的美感。与沙发相对应的白色椅子以及落地灯拥有极富设计感的造型,其与黑色系家具形成戏剧化的"对峙"。主人还在与厨房处于同一动线上的区域,设置了一个书房。阅读与美食空间的组合也形成了一种默契,这种大胆而有趣的设计,像极了好莱坞影片中的场景,幽默但不脱离实际。

后现代新古典

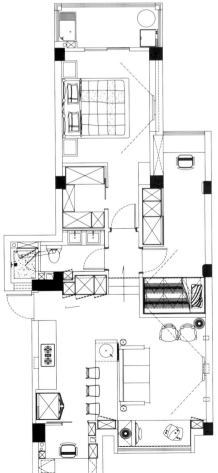

For the start of the design, the host branded some distinct material label for this new home, which is "No ceramic tile is needed." Upon entering the space, "cool" is the best word which can best interpret the regions that the eyesight catches. With fewer colors, accompanied with rigidness created by metal and glass, the masculinity of the property owner is produced.

For this project, the living room and the kitchen has open layout. Bar counter is the media connecting these two functional areas. The texture of the bar counter is unified with the ground and the TV background wall. These mottled and antique outlooks carry some industrialized hints. This kind of vicissitude temperament reveals some irresistible mysterious sensations. Naturally these areas are enchanting to every guest. The chairs of the bar counter create the atmosphere of the space with modern and concise appearance. The texture of the white metal duly mitigates the harshness of kitchen. Geometric lines are displayed inside the space. It is hard for us to classify this design into any specific style. And that might be out of the ingeniousness of the designer.

The living room next to the kitchen has black and white colors as the tone colors. The grandness of black leather sofa appears quite in place within the residential space with such atmosphere, with some solid aesthetic perceptions. The white chairs and the floor lamps corresponding with the sofa have formats with strong design feel, which form dramatic contrast with the black furniture. The host even set a study in an area which is at the same moving line with kitchen. There is some tacit agreement arising from the combination of reading and feast space. This bold and interesting design is quite like some scene from Hollywood movies, humorous but not isolated from reality.

宝安中洲中央公园样板房
Show Flat of Zhongzhou Central Park in Bao'an

设计单位：KSL 设计事务所	Design Company: KSL DESIGN(HK) LTD.
设 计 师：林冠成、温旭武、马海泽	Designers: Andy Lam, Wen Xuwu, Ma Huize
项目地点：广东省深圳市宝安区	Project Location: Bao'an District, Shenzhen City
项目面积：180 m²	Project Area: 180 m²
主要材料：手扫漆、木地板、皮板、皮革、玫瑰金装饰板	Major Materials: Lacquer, Wooden Flooring, Leather Board, Leather, Rose Gold Decorative Board

后现代新古典

如何让空间更为简洁、大气,线条更加唯美,家的温馨氛围更为浓烈?这是设计师一直以来在探索的设计问题。

在本案中,设计师以后现代简约风格为主,在家具以及其他细节上选用了一些奢华风格的具有新古典气息的线条凝练的装饰品。设计初衷就是使这个空间的每一个角落、每一处细节都更为精致,为业主打造出高品质的生活空间。

设计师将色彩定位为沉稳、内敛而雅致的灰咖色系,对于装饰画、绿植等细部用红色、墨绿色等亮色调打破咖色的沉稳,活跃了空间的氛围。在材料的使用上,地面大面积铺设木地板,墙面上使用咖色的皮板,卫浴间大面积使用茶色钢化玻璃。灯光与材质交相辉映,设计师用最简练的设计手法,实现了简约而不简单、奢华而含蓄的平衡。

How to make the space more concise, magnificent, lines nicer and the warm atmosphere of home more intensive? This is the design issue that designers have always been working on.

For this project, the designer focused on post-modern concise style and selected some luxurious ornaments with concise lines and neo-classical atmosphere for furniture and other detail parts. The designer intends to make every corner, every detail much more exquisite and create high-quality life space for property owners.

The designer selects grey coffee color as the tone color which is sedate, restrained and elegant. For detail parts such as decorative painting, green plants, etc., the designer selects bright color tones such as red, dark green to break the sedateness of coffee color and activate the atmosphere of the space. As for the application of materials, the ground is paved with wood flooring for a large area. The wall applies coffee color leather board, and the washroom is applied with tawny toughened glass. The lights and the materials bring out the shining in each other. With most concise design approaches, the designer achieves concise but not simple, luxurious but implicit balance.

保利叶之语样板房
Poly Yezhiyu Show Flat

设计单位：上海乐尚装饰设计工程有限公司
设计师：卢尚峰
项目面积：82 m²

Design Company: Shanghai Lestyle Decorative Design and Engineering Co., Ltd.
Designer: Lu Shangfeng
Project Area: 82 m²

后现代新古典

在本套样板房的设计中，我们根据人物性格、年龄、社会地位、喜好等因素，将之设定为新古典风格的样板房设计。

一个世纪以来，新古典风格横穿欧亚大陆，无论新派风格如何兴起，新古典风格总是以低调沉稳的精神内涵散发出高贵的气势，超越了国界与地域的限制。它有一种醇厚的形式美，摒弃了繁复的装饰，以最经典的古典元素，最简约的表现手法，展现历史与文化纵深感。另外，新古典更加强调人与人、人与社会之间的关系，把单一的空间，变成一种富有人情味的生活场景。把与社会和自然相隔绝的空间，变为一种同社会和自然展开对话的空间。

For the design of this show flat, the designer set it to be show flat design with neo-classical style based on the elements of human characteristics, age, social status, likes, etc.

For over a century, neo-classical style spreads all across European and Asian continents. No matter how many new styles there are, neo-classical style would always send out some noble momentum with some low-key and sedate spiritual connotations, surpassing the limitation of regional and territorial boundaries. It has some mellow formal beauty, forsaking complicated decorations, but displaying historical and cultural vertical feeling with the most classical elements and most concise representation approaches. Other than that, neo-classical style more emphasizes on among people, between people and the society. This monotonous space is changed into some life scene full of human kindness. This space secluding people from the society and nature is changed into a space which can have a dialogue with the society and nature.

宁波镇海维科样板房
Show Flat of Weike Property in Zhenhai of Ningbo City

设计单位：上海大匀设计中心	Design Company: Shanghai Symmetry Space Design
设 计 师：陈雯婧、郭岚	Designers: Chen Wenjing, Guo Lan
软装设计：上海太舍馆贸易有限公司	Soft Decoration Design: Shanghai MOGADECO Co., Ltd.
软装设计师：汪晓理、李璐璐	Soft Decoration Desginer: Wang Xiaoli, Li Lulu
项目面积：104 m²	Project Area: 104 m²
项目地点：浙江省宁波市	Project Location: Ningbo in Zhejiang Province
主要材料：竹木染色、地毯、枫桦木地板	Major Materials: Dyed Bamboo Wood, Carpet, Birch Wood Flooring

后现代新古典

住宅设计是人类创造更好的生存和生活环境的重要活动，它运用现代化的设计原理打造出舒适、美观的设计，使空间更加符合人们的生理和心理需求。在空间的划分上，设计师也考虑到了这一点。

本案中，设计师将材料进行综合运用，讲求空间的几何表情和功能划分，用皮革和铸铁打造出具有浓郁装饰感的墙面。沉静的棕米色系以及布满装饰花纹的布面，呈现出轻快、活泼的流畅之美，缔造出不一样的空间情调。

Residential design is some important social activity for people to create better surviving and living environment. It applies modern design approaches to produce comfortable and nice design, thus making the space more appropriate for people's physical and psychological requirements. The designer also takes that into consideration for the aspect of space division.

For this project, the designer makes comprehensive use of the materials and focuses on the geometric expression and functional division of the space, creating a wall with intensive decorative feel with leather and cast iron. Serene brown color and cloth of decorative patterns create dynamic and brisk beauty, creating some different space appeals.

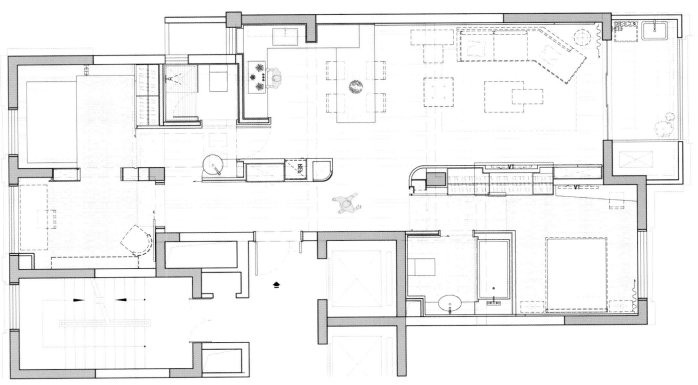

爱尚摩登 Fashion Modern

设计单位：DOLONG 设计
施工单位：大品专业施工
项目面积：141 m²
主要材料：简一大理石、仿古地板、拼花陶瓷锦砖、软包、数码彩油漆
摄　　影：金啸文空间摄影

Design Company: DOLONG Design
Construction Company: Dapin Professional Engineering
Project Area: 141 m²
Photographer: Jin Xiaowen Space Photography
Major Materials: Marble, Antique Floor Board, Ceramic Mosaic Tile, Soft Roll, Digital Colorful Paint

优雅利落、品位不凡，是这个空间给人的视觉印象。设计师用协调的色彩、清新简约的线条，勾勒出一个个生动的空间画面，无论从哪个视角观看，空间的美感与质感皆值得称道。

客厅的基调利落明快，高贵的黑色与纯净的白色相互映衬，使得空间色彩的节奏活跃起来。为避免空间氛围过于沉闷，特加入红色亮片的靠枕。背景墙上那翩翩起舞的蝴蝶，使得整体空间看起来柔和了不少，弥补了空间的冰冷与单调。简洁的线条、黑与白为主的色彩，偶有温情的原色掺杂其中，使得整个空间呈现出柔媚的气质。

餐、厨空间合二为一，使得空间功能得到了最大程度的发挥。厨房大量运用白色作为主打色，且恰如其分地点缀了稍许黑色，展现出简洁干练的气质。白色的餐桌、餐椅的设计非常简洁，具有时尚感。

卧室里窗户面积很大，把充足的光线和温暖的阳光聚集起来，空气中弥漫着淡淡的清香味。一张黑色的大床设计简约而精美，大大的尺寸几乎占据了整个卧室。干净的白色家具搭配稳重的咖啡色窗帘，使卧室看起来理性、含蓄、现代味十足。

This space leaves people with the visual impression of being elegant, clear and high taste. The designer makes use of harmonious colors, fresh and concise lines to create one and after after vivid space pictures. No matter from what view angle, this space attains high aesthetic level and texture.

The tone of the living room is clear and brisk. Noble black colors correspond with clear white, activating the color rhythm of the space. In order that the space atmosphere does not appear much too dumb, the designer adds some cushion with red paillettes. The dancing butterflies on the background wall soften the whole atmosphere of the space, remedying the coolness and monotonousness of the space. The whole space displays some gentle and lovely temperament with concise lines, black and white colors and softhearted colors.

The designer combines the dining restaurant and the kitchen, bringing about the best expression of the space functions. The kitchen makes a lot of white as tone color, appropriately dotted with some black colors, displaying some concise and brisk temperament. The white dining table and the chairs are quite simple in design, full of fashion feel.

The bedroom has large windows, integrating sufficient lights and warm sunshine, making the space full of light fragrance. A black bed is simple but elegant, with its large size taking the whole area of the bedroom. The clear white furniture is allocated with sedate coffee color curtain, making the whole bedroom appear reasonable, low-key and modern.

逸静新贵主题样板间
YiJing Neo-Aristocracy Show Flat

设计单位：J2-DESIGN/厚华装饰设计有限公司
设 计 师：卓晓初
陈设与选材设计师：谭启开
开 发 商：广东新中源集团
项目地点：广东省肇庆市
项目面积：115 m²
主要材料：硬色壁纸、灰镜、灰钢、陶瓷锦砖

Design Company: J2-DESIGN/Houhua Consultant Design Co.,Ltd
Designer: Zhuo Xiaochu
Layout and Material Designer: Tan Qikai
Developer: New Zhongyuan Group in Guangdong Province
Project Location: Zhaoqing in Guangdong Province
Project Area: 115 m²
Major Materials: Hard Color Wallpaper, Grey Mirror, Grey Steel, Ceramic Mosaic Tile

后现代新古典

本案为现代港式风格，流露出浓浓的新古典情调和氛围。整体色调以米黄色为主、咖啡色为辅。设计师运用长条块面相拼接的手法，将不同的材料结合在一起。简练的横条相错拼，表现出块面感十足的现代家居的豪华感，简洁而不失时尚与优雅之感。

过道的中央位置，设计师将灰钢和灰镜错拼在一起，与地面上的陶瓷锦砖相呼应，成为整个设计的亮点，起到了连接客厅和卧室的作用。墙面采用具有凹凸肌理的两种颜色的壁纸，显示出丰富而有序的肌理纹样，材料的相互贯穿使得客厅与卧室相互连接起来，形成了整体而有序的空间氛围。

This case has modern Hong Kong Style, displaying intensive neo-classical charms and atmosphere. The whole color tone is beige-based and supplemented by brown color. The designer makes use of the collage of strips and chunks to integrate different materials. Concise horizontal strips are pieced together in staggered arrangement, displaying luxurious architecture of modern furnishing with strong sense of chunks, which makes the whole space fashionable and elegant.

In the central area of the corridor, grey steel and grey mirror are placed together by the designer, corresponding with the ceramic mosaic tile on the ground which becomes a highlight for the whole design and functions as the connection between the living room and the bedroom. The wall makes use of two colors of wallpaper with concave-convex texture, displaying rich but orderly pattern. The inter-fingering of materials connects the living room with the bedroom, producing whole and orderly space atmosphere.

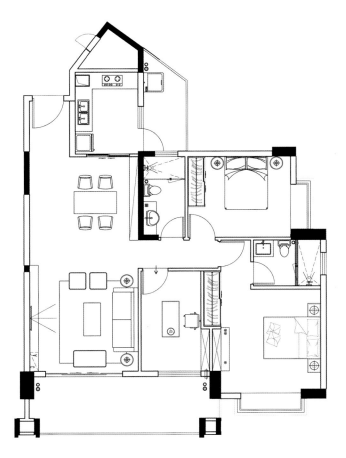

镇江优山美地

Zhenjiang Youshan Meidi Property

设计单位：TY34 精品设计中心
主设计师：庄光科
软装设计师：马翠琴
项目面积：200 m²
主要材料：陶瓷锦砖、爵士白大理石、意大利灰大理石、水晶瓷砖
摄影师：金啸文

Design Company: TY34 Boutique Design Center
Leading Designer: Zhuang Guangke
Soft Decoration Designer: Ma Cuiqin
Project Area: 200 m²
Major Materials: Ceramic Mosaic Tile, Jazz White Marble, Italian Grey Marble, Crystal Ceramic Tile
Photographer: Jin Xiaowen

优山美地位于镇江京口区谷阳路和禹山北路交汇处,总占地面积37万平方米,绿化率43%。规划有别墅、洋房、小高层、高层、酒店式公寓等多种业态产品。

在本案中,设计师结合得天独厚的自然优势,将本案的室内与室外景观融合起来。大面积的开窗搭配素雅的双层窗帘,将室外景观很自然的引入室内,为时尚现代的居住空间带来了一份大自然的气息。

在材料运用上,设计师大量使用现代气息浓厚的瓷砖、大理石类石材,营造出干练简洁的居住环境。在软装饰的应用上,橘色的沙发、黑白红色块相加的地毯、艺术气息浓厚的墙饰,都为简练的空间环境增加了一分温柔的气息。

This project is located in the junction of Guyang Road and North Yushan Road in Jingkou District of Zhenjiang, with total floor area of 370,000 square meters and greening rate of 43%. According to the planning, there are various kinds of spaces, such as villa, foreign-style house, small high-rise, high-rise and service apartment.

For this project, the designer combines the advantaged natural circumstances and integrates interior space with the exterior landscape. Large area of window match elegant curtain of two layers, naturally introducing the outdoor landscape inside, bringing some natural atmosphere for the fashionable and modern residential space.

For the application of materials, the designer makes a lot of use of ceramic tiles, marble, etc. of intensive modern smell, displaying a concise and clear residential space. For the application of soft decorations, orange sofa, carpet of black, white and red color blocks and wall ornaments of artistic sphere all add some gentle atmosphere foe this concise space environment.

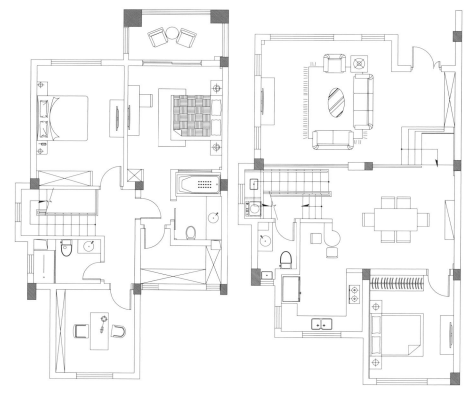

静聆风吟·新东方寓所空间
Listen to the Chanting of Wind · New Oriental Residential Space

设计单位：福州创意未来装饰设计有限公司
设 计 师：郑杨辉
项目地点：上海市
项目面积：180 m²
主要材料：方钢、浅灰色玻化砖、得高软木

Design Company: Fuzhou Creative Future Decoration and Design Co., Ltd.
Designer: Zheng Yanghui
Project Location: Shanghai
Project Area: 180 m²
Major Materials: Square Steel, Light Grey Vitrified Brick, Cork Wood

每个人的心里都装着一个关于家的梦想，看着城市中亮起的万家灯火，只有温暖的家最贴近我们的心灵。"静聆风吟"是一位儒雅的成功人士对自己寓所的期盼，因而这所新东方的寓所中所散发出的内敛、尊贵、淡定、从容的空间气质就是设计师所要表达的空间精神。

在平面动线的规划上，设计师将原有户型中的入户区划归为餐厅空间，做到餐厅与厨房空间的直接互动，同时也能够引入室外的光线，并且通风良好。客厅区域和半敞开的书房空间最大限度地塑造了与家人沟通互动的空间场景。

在空间光效的表达上，设计师选用了暖色的灯光。空间中基本上都使用了直线的造型，大面积的材质应用富有整体性，在材质的变化中又体现出空间的多样性。在富有肌理的材料质感和色彩的协调下，完美的空间骨架就这样构建出来。

Inside everyone's heart, there is a dream about home. Looking at the lamps and candles of myriad families inside the city, our heart can only feel close to a warm home. This project is the expectation of a learned and refined successful people towards his residence. Thus, the restrained, noble, calm and deliberate space temperament that this new oriental residence sends out is the space ethos that the designer intends to express.

For the plan planning, the designer changes the entrance area of the original flat into dining space, thus forming direct interaction between dining hall and kitchen. While exterior light can be brought inside and fine ventilation environment is created. The living room and the semi-open study maximize the space for family members to communicate with each other.

For expression of space lighting effect, the designer selects warm lights. Basically the space makes use of straight formats. There is some integrity in the large area application of materials. The diversity of space is displayed in the variation of materials. Perfect space framework is thus constructed with the coordination of texture of materials and colors.

设计单位：鼎汉唐设计机构
设计师：陈明晨、陈墩华、沈江华
项目地点：福建省福州市
项目面积：310 m²
主要材料：新米黄大理石、实木地板、蒙托漆、壁纸

Design Company: Design Hand Institution
Designers: Chen Mingchen, Chen Dunhua, Shen Jianghua
Project Location: Fuzhou in Fujian Province
Project Area: 310 m²
Major Materials: New Cream-Colored Marble, Solid Wood Flooring, Paint, Wallpaper

波光艳影

这是一个简约的纯净世界，它简洁大方，灵动非凡，非常适合追求低调、典雅生活氛围的业主。设计师将简洁的设计手法巧妙地应用于雍容华贵的空间中，营造了一个大气非凡的经典空间。

客厅部分最吸引人眼球的是电视背景墙的木质隔断与楼梯，这也是设计师最为得意的地方。由特制的饰面板拼接而成的空间拥有了更多的结构感和现代感。和谐的对比存在于空间中，简洁的白色吊灯和大理石地面交相辉映，白色的瓷质餐具、透明优雅的高脚酒杯与深褐色的餐具、暖棕色的餐椅形成强烈的视觉对比。但是在休闲区域却又恰恰相反，黑色地毯覆盖着地面，反衬着米色的布艺沙发，给人以洁净、舒适之感。

装修中所涉及的一些内容，如饰面板、沙发的造型及颜色、饰品及灯具造型、玻璃护栏的高度、卫生间洁具的选择，甚至室内绿色植物的品种以及摆放位置，都是设计师经过精心构思的结果。这种构思使得空间温馨而富有质感，充满了雅致的情趣。

Glistening Waves and Colorful Shadows

This is a simple and pure world which is generous, dynamic and uncommon and which fits for property owners aspiring for low-key and elegant life atmosphere. The designer ingeniously applies concise design approaches into this noble and aristocratic space, creating a magnificent classical space.

The most attractive part inside the living room is the wood partition and staircase of the TV background wall which the designer is mostly proud of. The space made of tailor-made veneer possesses many structural feel and modern feel. Harmonious contrast exists in the space. Concise white droplights correspond with the marble ground. White porcelain tableware and elegant goblets perform intensive visual contrast with the dark brown tableware and brown dining chairs. Things are quite the opposite for the leisure space. The floor is covered with black carpet, setting off the beige cloth sofa, making people feel relaxed.

Many aspects that the decoration is concerned with, such as veneer, modeling and color of sofa, modeling of ornaments and lighting accessories, height of glass guardrails, sanitary appliance of the washroom, even the species of green plants and locations, all these are the outcome of careful conception by the designer. This kind of conception makes the space warm and full of texture, with elegant interests.

中庚城某住宅 One Residence inside Zhonggengcheng Property

设计单位：福建国广一叶建筑装饰设计工程有限公司	Design Company: Fujian Guoguangyiye Construction Decorative Design and Engineering Co., Ltd.
方案审定：叶斌	Project Examiner: Ye Bin
设 计 师：张晓玲	Designer: Zhang Xiaoling
项目地点：福建省福州市	Project Location: Fuzhou in Fujian Province
项目面积：130 m²	Project Area: 130 m²
主要材料：石材、实木地板、壁纸、灰镜	Major Materials: Stone, Solid Wood Flooring, Wallpaper, Grey Mirror

后现代新古典

本案在做空间布局规划时，充分考虑业主的需求，使原本面积不大的空间呈现出延伸感，同时运用材料、家具、装饰品、照明等，营造出现代空间的时尚感。在设计手法上，设计师综合利用材料的光感效果，将空间的各个层次相互叠合，显示出不同空间的隐性分隔。

同时，设计师运用了多种材料，如石材、壁纸、木材、镜子等，通过其质感以及各自不同的特性，使这个空间在简约中散发出时尚的品位。观景阳台的一角栽植了一些饶有趣味的绿色植物，展现出妙趣横生的自然气质。

While composing space layout planning for this project, the designer takes the requirements of property owners into consideration and creates some extension for the original small space. While at the same time, the designer creates the fashion feel of modern space with materials, furniture, decorative objects, illuminating supplies, etc. For the design approach, the designer makes comprehensive use of light sensation of materials to create interactive superimposition of various levels which displays the recessive separation among different spaces.

While at the same time, the designer makes use of multiple materials, such as stone, wallpaper, wood, mirror, etc. to make this concise space send out some fashion taste through the texture and different materials of different materials.

设计单位：福建国广一叶建筑装饰设计工程有限公司
方案审定：叶斌
设 计 师：陈家雄
项目面积：110 m²
项目地点：福建省福州市
主要材料：金刚板、水泥漆、仿古砖

Design Company: Fujian Guoguangyiye Construction Decorative Design and Engineering Co., Ltd.
Project Examiner: Ye Bin
Designer: Chen Jiaxiong
Project Area: 110 m²
Project Location: Fuzhou in Fujian Province
Major Materials: Laminate Flooring, Cement Paint, Antique Brick

"大道至简,为画者善于留白。"

本案的一切设计从简出发,没有使用过多的装饰。在保留原始空间结构的基础上,设计师以中轴线来划分空间。通过灯光氛围的营造把空间做出明、暗两种层次;通过用阴阳角的收口将墙面和吊顶完美地结合在特定的空间结构里。

在色彩的搭配上,公共空间处以黑、白两色为主基调,咖啡色的家具实现了空间中黑与白对比的协调质感。在卧室的部分则以暖色调为主,使得空间连贯起来,并且还具有延伸性。餐厅中,黑色的吊灯和餐桌完美地融合在一起,这一组家具的组合也使得空间中黑白的对比更加富有层次。

简,即简练、简单。如一种减法生活,使人们回归到简单的生活中。

Great theories are all very simple. And great painters are good at leaving some blanks.

The design of this project starts from simplicity and does not use superfluous decorations. Based on maintaining the original space structure, the designer uses central axis to divide the space. Through the creation of light atmosphere, the space is divided into two layers, dark and bright. The Yin and Yang Corners design perfectly combine wall and ceiling into specific space structure.

For color collocations, the public space has black and white as the tone colors. The brown furniture achieves the harmonious texture of black and white contrast inside the space. The bedroom is focused on warm colors to connect the space, with extensive feature as well. In the dining hall, black drop lights integrate with dining table perfectly. The combination of this set of furniture make the contrast of black and white inside the space with richer layers.

Simplicity means succinct and simple. Just like some subtraction life which brings people back to the original simple life.

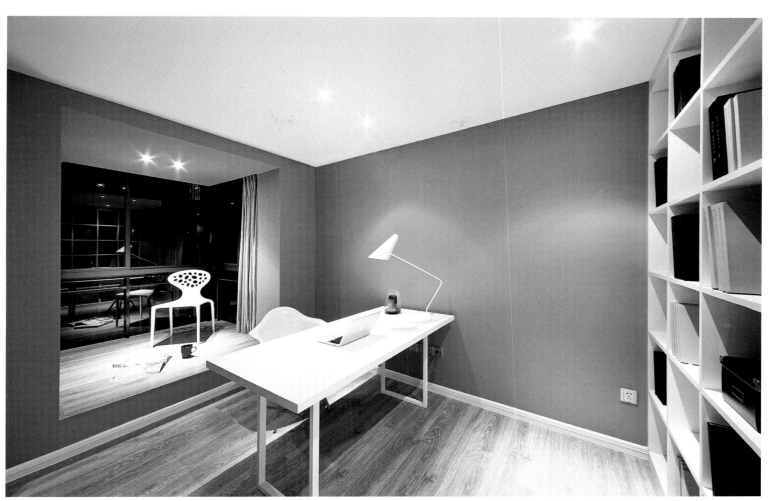

设计单位：北岩设计	Design Company: Beiyan Design
项目地点：江苏省南京市	Project Location: Nanjing in Jiangsu Province
项目面积：113 m²	Project Area: 113 m²
主要材料：艺术砖、硅藻泥、壁纸	Major Materials: Wallpaper, Diatom Ooze, Artistic Brick
摄　　影：金啸文空间摄影	Photographer: Jin Xiaowen Space Photography

后现代新古典

摩登样板间 II

这是一个三室两厅两卫的户型，比较方正，各个房间的采光也不错。房子是老人居住，他们对房间的储藏功能有更高的要求。两位老人接受新事物的能力比较强，不喜欢传统老人房设计中常用的老式家具以及暗色调。

经过相互的沟通，我们在空间布局上做了些许调整，把主卧室的卫生间改为衣帽间和储藏间，这样就增加了储藏功能。在业主的要求下，将主卧的飘窗拆除，放上一个摇椅，每天老人可以在阳光的沐浴下，品一杯浓茶，阅一份报纸，享受惬意的生活。书房门被改成宽阔的玻璃移门，左右比较通透，使得房间更加明亮，宽大的书桌是挥洒笔墨的好地方。主卧原房门的位置被改造成了洗面区，干湿分离的格局更加符合现代人的使用习惯。

在材质使用上采用了现代主义的装饰手法，墙面大面积采用艺术砖及硅藻泥，大气简约，而且更加符合环保的要求，以此营造出现代而温馨的家居空间。

This is a square flat with three bedrooms, two living rooms and two bathrooms, with each room having fining lighting. This house is owned by elder people who have higher requirements for the storage functions of the rooms. These two elder people have fine capabilities of accepting new things. They do not like the common old furniture and the dark color tone in the traditional houses for elder people.

After mutual communications, we make some adjustments towards the space layout, changing the washroom of the master bedroom into cloak room and storing space which creates some storage functions. Upon requirements of the property owner, the window of the master bedroom was removed to set aside space for a rocking chair. Thus the elder people can have a cup of strong tea and read a piece of newspaper bathed in the sun. The door of the study is changed into a broad glass sliding door with transparent space on both sides, making the room quite bright. The broad table is a perfect place for practicing painting. The place of the master bedroom's door is changed into wash basin area. This patter with clear distinction between dry area and wet area corresponds with modern people's habits.

For materials application, the designer applies the modernistic decorative approach, with large area of artistic bricks and diatom ooze on the wall, which is magnificent and concise. While this is quite in accordance with the ecological requirements, the designer creates some modern and warm residential space.

设计单位：威利斯设计有限公司	Design Company: Willis Design
设 计 师：巫小伟	Designer: Wu Xiaowei
项目地点：江苏省常熟市	Project Location: Changshu in Jiangsu Province
项目面积：90 m²	Project Area: 90 m²
主要材料：壁纸、皮质硬包、明镜、饰面板、仿木纹砖、陶瓷锦砖、实木地板	Major Materials: Wallpaper, Leather Hard Roll, Bright Mirror, Veneer, Imitation Wood Grain Brick, Ceramic Mosaic Tile, Solid Wood Floor

本案是一个婚房的设计，与以往人们印象中大红大紫、色彩浓郁的婚房设计有所不同，而是以黑、白、灰为主，局部点缀比较鲜艳的色彩。在设计的手法上，空间线条的处理干净而利落，不同材质的灵活搭配使得居室中冷艳的奢华感与家的温馨感协调起来。

此居室是一个小三居的户型，缺少了一个宽敞舒服的餐厅和一个功能齐全的主卫。设计师在做设计的时候，充分利用了原有空间，将这两点遗憾降到最低。必要的墙体改动，解决了餐桌的摆放问题，与酒柜相连接，不会使其显得突兀，同时满足三代人一起用餐的需求。主卧的内卫是由南北两个房间压缩而来，考虑到公寓房内排污的不便利，只安置了洗漱及冲淋的功能。

儿童房的设计中，设计师运用了榻榻米式的整体设计，这种设计的最大优点就是可以留出足够大的空间，作为亲子互动区。同时又将整体衣柜和书架都安置进来，学习功能与生活功能一应俱全。平时也可以作为喝茶、聊天、会友的休闲空间，享受阳光与宁静。

This is a design for the wedding room. This project is different from the normal wedding room design with intensive bright colors, but with black, white and grey as the color tone, dotted with some bright colors on some spots. For the design approach, the treatment towards the space lines is clear and tidy. The flexible collocation of different materials makes the quiet and magnificent luxury feel of the residence coordinate with the warm feelings of a home.

This is a small flat with three bedrooms, without a broad and cozy dining hall and a master bathroom with complete functions. While carrying out the design, the designer makes full use of the original space and gets

rid of these two shortages. The necessary transformation towards the wall resolves the arrangement issue of the dining table. The connection with the wine cupboard makes the space does not appear much too strange, while making it possible to accommodate three generations to have meals together. The inner bathroom of the master bedroom is condensed from both rooms at south and north. Considering the inconvenience for sewage disposal inside the apartment, there are only washbasin and sprinkler inside.

The designer applies tatami style whole design for the children's room. The biggest advantage for this design is that there can be sufficient room for interaction between parents and children. There are also wardrobe and book shelf here with complete functions for study and living. During daily life, this leisure space can be used for drinking tea, chatting and meeting friends, full of sunshine and tranquility.

九龙仓时代晶科名苑样板房
Wharf Holdings Shidai Jingke Mingyuan Show Flat

设计单位：多维设计事务所
软装设计：成都诺特软装饰工程有限公司
设 计 师：范斌、张晓莹
项目面积：65.8 m²
项目地点：四川省成都市
主要材料：白宫米黄、宝加莉灰、银白龙、石材陶瓷锦砖、壁纸、地毯、不锈钢、茶镜

Design Company: DODOV Design Studio
Soft Decoration Designer: Chengdu Nuote Soft Decoration and Engineering Co., Ltd.
Designers: Fan Bin, Zhang Xiaoying
Project Area: 65.8 m²
Project Location: Chengdu in Sichuan Province
Major Materials: Beige Marble, Silver Dragon Marble, Mosaic Stone, Wallpaper, Carpet, Stainless Stone, Tawny Glass

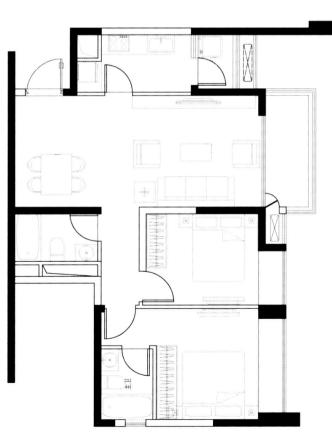

后现代新古典

本户型面积不大，居住面积为 65.8 m²，目标客户群主要为周边大型企业的中、高级员工。设计师通过局部空间叠合的设计手法，提高了空间的利用率，把 65.8 m² 的原空间改造为实际利用面积为 80 m² 的居住空间。

在设计定位上，设计师以现代、简洁为主格调。结合软装配饰的合理搭配，整体氛围和谐而统一，从而营造出舒适、时尚的个性生活空间。

色调整体上以灰色、米白色、浅咖色为主，白色、金色等作为点缀色。家具多采用中性色，具有质感的布料搭配少量的金属饰品，使软装与硬装相结合，在营造舒适氛围的同时也体现了时尚感。

在灯光效果的营造上，设计师通过台灯、落地灯等辅助光源，营造出丰富的灯光层次，体现出现代及温馨的感觉。整个氛围突显出主人内敛、高贵的内在气质。

The area of this type of flat is not big, with living area of 65.8 m². This project is mainly oriented for medium and advanced-level staff of grand corporations around. Through superimposition design approach of partial space, the designer enhances the use ratio of space and changes the original 65.8 m² space into residential space with practical usable area of 80 m².

As for design orientation, the designer takes modernity and conciseness as the key tone. Combined with the appropriate collocation of soft decoration ornaments, the whole atmosphere appears harmonious and consistent, thus creating comfortable and fashionable unique living residence.
\

The color tone is focused on grey, creamy white and light coffee colors, with white and golden colors as the ornamental colors. Most furniture has neutral colors. Cloth of texture is combined with some metal ornaments, integrating soft decoration with hard decoration, which creates some comfortable atmosphere and presents some fashionable feel.

For the creation of lighting effect, through secondary light sources such as table lamps and floor lamps, the designer creates lights with abundant layers, displaying modern and warm sensations. The whole atmosphere highlights the restrained and noble inner temperament of the property owner.

凯德置地御金沙交楼标准样板房
Yujinsha Show Flat · Kaide Land Holdings

设计单位：广州共生形态工程设计有限公司
WWW.COCOPRO.CN
项目地点：广东省广州市金沙洲

Design Company: Guangzhou C&C Design Co., Ltd.
WWW.COCOPRO.CN
Project Location: Jinshazhou in Guangzhou, Guangdong Province

本案空间布局紧凑，设计风格俏皮、雅致、唯美。客厅中，风铃状的水晶吊灯摇曳多姿，沙发墙面上的方形体拼贴出的印象风格的装饰画高雅、精致，与餐厅墙面上用纽扣拼贴出的装饰图案遥遥相对，显示出居室主人不落俗套的欣赏品味。地面上的酒红色绒毛地毯为空间增添了些许高贵的气质。其他的一些时尚的摆设使空间多了一份前卫的现代感，小孩房中活泼的色彩表现了孩童的天真与纯洁，千纸鹤造型的吊灯形态优雅，形态各异，显示出强烈的创造性。书房中高低错落的书柜配合了整个空间节奏，体现出现代空间的不落俗套感。

The space layout of this project is terse, with smart, elegant and aesthetic design style. Inside the living room, the crystal chandelier shaped like wind chime sways with elegance. The decorative painting of impression style pasted with square objects on the wall of the sofa presents elegance and exquisiteness. This painting echoes the decorative diagram made of fasteners on the wall of the dining hall, displaying the uncommon appreciation taste of the property owner. The red wine color carpet on the floor adds some aristocratic temperament for the space. Other fashionable decorative objects create some avant-garde modern feeling for the space. The dynamic colors of the children's room display the naivety and innocence of children. The chandelier shaped like paper crane looks graceful and has various formats, displaying intensive creativity. The high and low bookcases inside the study match the rhythm of the whole space, displaying the uncommonness of modern space.

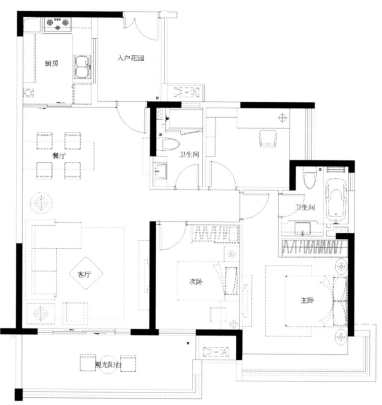

香榭丽舍住宅公寓
Champs Elysees Residential Apartment

设计单位：林卫平室内设计事务所
设计师：林卫平
项目面积：180 m²
项目地点：浙江省宁波市
主要材料：爵士白、镜面玻璃、壁纸饰面

Design Company: Lin Weiping Interior Design Firm
Designer: Lin Weiping
Project Area: 180 m²
Project Location: Ningbo in Zhejiang Province
Major Materials: Jazz White Marble, Mirror Glass, Wallpaper Veneer

后现代新古典

色彩交响曲

色彩是空间的音符，最能感染人的情绪。在这套香榭丽舍的作品中，蓝、白、红、紫四色构成了空间的主旋律。明快的蓝色以及白色奠定了客厅及公共空间轻松的基调，并给人理智、平静、清新的感受；而红色则让餐厅的氛围显得活泼，卧室的气氛变得浪漫、神秘；尊贵的紫色总是给人矛盾的美感：一方面，它是沉静的，另一方面，它又让人有梦幻的感觉。

颜色的交汇演变成一首动人的交响曲，给人们带来了精神上的愉悦，大大提高了空间与人的情感沟通。设计师结合形态、材质、装饰等设计要素和虚实相生的设计手法，使得空间在简约的基调中衍生出无数的变化，充满迷人的氛围和浪漫的情趣。

本案中，简洁的家具风格、丰富的色彩变化与整个空间遥相呼应。在尊重空间的同时，整个设计赋予空间一丝灵动之美。空间充满知性、童趣、跃动与脱俗的氛围。即使不经意一低首，也能发现惊喜与可爱的地方。

Symphony of Colors

Colors are the musical notes of the space which is most infectious to people's sentiments. For this work of Champs Elysees, Blue, White, red and purples colors compose the mainstream of the space. Brisk blue and white become the relaxing tone for living room and public space, giving people reasonable, tranquil and fresh sensations. Red makes the atmosphere of the dining hall appear lively and active and the bedroom romantic and mysterious. Noble purple colors would always leave people with contradictory aesthetic feelings: on one side, it is tranquil, and on the other side, it gives people dreamlike sensations.

The convergence of colors becomes a touching symphony, creating great spiritual pleasures for people and greatly uplifting the emotional communication between the space and people. The designer combines design elements such as formats, materials, decorations and the design approach integrating false and true, creating limitless variations from the concise tones, with enchanting atmosphere and romantic sentiments.

For this project, concise furniture style and rich color variations echo the whole space. On respecting the space, the whole design endows the space with some ethereal beauty. The space is filled with intellectual, interesting, lively and uncommon atmosphere. Even you look around casually, you can find some surprising and lovely parts.

江南人家 Jiangnan Residence

设计单位：易百装饰（新加坡）集团有限公司
设 计 师：冯易进
项目面积：135 m²
主要材料：微精石、饰面板、硬包、壁纸

Design Company: E&B decoration (Singapore) Group Co.,Ltd.
Designer: Feng Yijin
Project Area: 135 m²
Major Materials: Microlite, Veneer, Hard Roll, Wallpaper

设计师在本案中没有花更多的时间去营造某一种特定的风格,而是理性地思考如何在空间中创造愉悦的氛围。

电视墙的材质从墙面延伸到地面,甚至到了鞋柜,同一种材料在同一个空间的延伸就形成了一个大面积空间的格局,使得两个空间界面在视觉上形成了一个有机的整体。

也许有人觉得地面和墙面的花纹略显花哨,对业主而言,这种石材的肌理以及花纹却是空间的一大亮点,于是我们把这个亮点放大,使之成为一个整体,反倒使空间有了绚烂的奢华感。

沙发背景墙应用了中间色的暗橙色的木板,中和了石材的这种绚烂,与橙色和米白色的沙发搭配协调。本案没有过多的装饰品,却在简单的氛围中营造出了温馨的情调。

For this project, the designer does not spend a lot of time creating some particular style, but thinks reasonably about how to create some pleasing atmosphere in the space.

The materials of the TV background wall extends from the wall towards the ground, even to the shoe cabinet. The expansion of the same materials inside the same space produces some grand space state, visually connecting these two spaces and producing an organic unity.

Some people may think that the patterns of the ground and wall appear a little bit gaudy. But for the property owner, the texture of the stone and the pattern is a highlight in the space. Thus we enlarge this bright spot and make it become a whole, creating some gorgeous luxury feel inside the space.

The sofa background wall applies dark orange color wood board, softening the color of the stone and coordinating in harmony with the orange and beige white sofa. There are no excessive decorative objects, but with warm tones inside this simple atmosphere.

光与影的舞蹈
Dance of Light and Shadow

设计单位：福建国广一叶建筑装饰设计工程有限公司	Design Company: Fujian Guoguangyiye Construction Decorative Design and Engineering Co., Ltd.
方案审定：叶斌	Project Examiner: Ye Bin
设 计 师：林庆华	Designer: Lin Qinghua
主要材料：枫木、玻璃、壁纸、陶瓷锦砖、雪弗板	Major Materials: Maple Wood, Glass, Wallpaper, Ceramic Mosaic Tile, PVC Expansion Sheet
摄 影 师：施凯	Photographer: Shi Kai

本案的设计注重和谐氛围的营造以及写意气质的书写。在纯净的光与影交汇的空间中，各个空间相互渗透又相互独立，营造出惬意、休闲的生活场景。零星点缀着的色彩也相映成趣，极大地契合了主人淡然的品性。

客厅中深、浅色调婉约流畅，电视墙面上的半透明玻璃映衬出枫木饰面板的纹理，显得趣味十足。在质感硬朗的空间中，地面上的圆形翻毛地毯为空间增添了一丝舒缓的气息。居室中，最有意思的设计还是雪弗板雕刻的镂空图案的隔断和门板。书房中、走廊里，影影绰绰的曼妙光影充盈在空中，漫步在此，就像徜徉于洒满阳光的花园里，给人以无限的美感。

The design of this project focuses on the creation of harmonious atmosphere and poetic temperament. In this pure space integrating light and shadow, various spaces are mutually infiltrating and self independent, creating cozy and leisurely life atmosphere. The colors decorated in some fragmentary way bring out the best in each other, which tally with the calm characteristics of the property owner.

The dark and light colors of the living room are graceful and fluent. The semi-transparent glass on the TV background wall set off the texture of the maple wood veneer, full of fun. Within this masculine space, the round carpet of frizzled feather on the ground adds some releasing atmosphere for the space. The most interesting design within the residence is the partition and door plank carved with PVC expansion sheet. In the study and the corridor, the lithe and graceful light and shadow fills the space. Walking here, it is like wandering the sunny garden, giving people limitless aesthetic feelings.

保利东语花园样板房·水韵
Poly Dongyu Garden Show Flat · Charms of Water

设计单位：广州道胜装饰设计有限公司
项目地点：广东省佛山市
项目面积：76 m²
主要材料：雅士白复合石、水曲柳饰面板、写真玻璃、ICI乳胶漆、陶瓷锦砖、银镜、镜钢

Design Company: Guangzhou Daosheng Decorative Design Co., Ltd.
Project Location: Foshan in Guangdong Province
Project Area: 76 m²
Major Materials: Jazz White Composite Stone, Ashtree Veneer, Glass, ICI Emulsion Paint, Ceramic Mosaic Tile, Silver Mirror, Mirror Steel

后现代新古典

摩登样板间 II

设计师运用人们亲水的本性，为这个坐落在水库旁的小户型样板间设定了主题——水韵。首先，设计师对整体空间布局进行了改善，原建筑中餐厅的采光不够，所以我们将厨房打通做成开放式，这样的设计能有效地利用阳光，同时也使得原本较小的空间在视觉上显得开阔起来。因为户型较小，所以没有单独设计书房，设计师将餐厅的一面墙作为书架，令餐厅同时兼具书房的功能，使空间更具生活韵味。设计师还将卫生间的外墙面安置了一整面镜子，将门隐藏于整面镜子中，这样的巧妙设计使得空间更开阔而整体。

因为整体采光度不够，所以在色调上我们选择最明亮的白色为主基调，搭配海洋的靛蓝，营造出一个海洋的世界，体现出年轻夫妇对海洋的喜爱。

客厅的墙面、卧室的衣柜门都以蓝色的水、蓝色的鱼作为装饰，使整个空间就像被海水围绕，加上挂画中的水母，随处可见的珊瑚、海星，无处不在的蓝色……在这个空间你能深刻地感受到自己置身于海洋中，这也正是设计师想要传递给大家的感受，对大海、对水的热爱之情。

The designer makes use of people's love for water and sets the theme of this small size show flat neighboring to water reservoir to be Charms of Water. In the first place, the designer improves the overall space layout. Based on that the lighting for the dining hall is not sufficient, thus the kitchen was made open. Hence the design can make effective use of sunshine and original small space appears broader. Due to that the show flat is kind of small, thus there is no study and the designer changed a wall inside the dining hall to be a book shelf, making the dining hall possess the functions of being a study, full of life taste. The designer also installed a full wall of mirror on the outer wall of the bathroom, hiding the door in the whole mirror, which makes the whole space appear expansive and wholistic.

Due to that the whole lighting is not enough, the designer chose bright white color as the tone color. Accompanied with azure ocean color, some ocean world is created, displaying the young couple's love for the ocean.

The wall of the living room and the wardrobe door all apply blue water and blue fish as the decorations, making the whole space surrounded with ocean water. With the jellyfish in the painting, coral, starfish and blue everywhere... This space would make you feel that you are among the ocean. And that is the sensation that the designer wants to convey to people, some passionate admiration for ocean and for water.

镇江科苑华庭
Zhenjiang Keyuan Mansion

设计单位：TY34 精品设计中心
主设计师：庄光科
软装设计师：马翠琴
项目面积：400 m²
主要材料：陶瓷锦砖、木纹砖、墙纸、饰面板
摄 影 师：金啸文

Design Company: TY34 Boutique Design Center
Leading Designer: Zhuang Guangke
Soft Decoration Designer: Ma Cuiqin
Project Area: 400 m²
Major Materials: Ceramic Mosaic Tile, Wood Grain Brick, Wallpaper, Veneer
Photographer: Jin Xiaowen

电视背景墙的设计非常有特色，设计师利用钢结构直接从顶面悬吊了一面墙体，中间镂空了不少面积，配上灯光，既可以充当客厅与餐厅之间的隔断，也可以悬挂客厅的电视，两者兼得。客厅茶几地面上的灯，看起来很随意，但是很有品位。

客、餐厅是一个开敞化的空间，用一个青蓝色陶瓷锦砖的隔断做分隔，左边是餐厅，右边是客厅。整个客、餐厅，都是以点光源为主，所有的主灯都是以装饰为主，照明为辅。

二楼的书房，连着一个阳台，两边是客房。镂空的隔断分隔了书房与走道。木地板和竖条的隔断互相呼应，是这个空间的亮点。

主卧室的床头背景墙用了有肌理花纹的PU软包，主卫和主卧之间用了纯玻璃的隔断，墙纸也很有质感。

地下室包括一间客房、一间活动室、一间影视厅。影视厅里面设计了一个很长的吧台和一整面墙的酒柜，很气派。

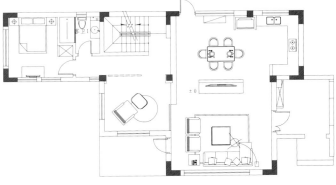

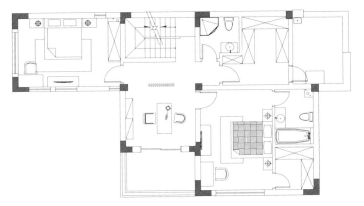

The TV background wall has very peculiar design. The designer hung a wall from the ceiling with steel structure with some hollowed space inside. Accompanied with the lights, this can not only be used as partition between the living room and the dining hall. The TV inside the living room can be hung here. The lights on the floor of the living room appears quite casual, with high taste.

The living room and dining hall is an open space, using ultramarine ceramic mosaic tile as the partition. On the left side is the dining hall and on the right the living room. The whole living room and the dining hall center on spot light source. All the major lights are focused on decorations with lighting as the auxiliary functions.

In the basement there is a guest room, an activity room and a movie theatre. Inside the movie theatre there is a long bar counter and a whole wall wine cabinet, quite magnificent.

The study on the second floor connects with a balcony, with guest rooms on both sides. The hollow-out partition separates the study from the corridor. The wood floor and the vertical partition correspond with each other, which is the light spot inside the space.

The bedside background wall of the master bedroom applies PU soft roll of texture pattern. The master washroom and master bedroom apply glass as the partition, and the wallpaper is of high texture.

微设计系列之暮光
Evening Twilight of Micro Design Series

设计单位：易百装饰（新加坡）国际有限公司
设 计 师：冯易进
项目面积：170 m²
主要材料：微晶石、饰面板、软包、壁纸

Design Company: Ebey Decorative Design (Singapore) International Co., Ltd.
Designer: Feng Yijin
Project Area: 170 m²
Major Materials: Microlite, Veneer, Soft Roll, Wallpaper

本案设计从一切微元素出发，不要假奢华、不要高调派、不要伪贵族，坚持没有风格的风格，所以称之为微设计系列……

在本案的色彩设计上，以高级灰为主，赭石、咖啡色为辅，避免出现多种色差。材质基本上以温馨感极强的布艺及板材为主。客厅部分的色彩分为三大块，灰色与赭石色用量相当，顶部以白色来过渡。造型上也以简约的微设计手法，来体现大与小、多与少的对比与变化。

电视墙上应用了大量统一的材质——软包，虽然是常用的材质，却在不同的灯光渲染下，以小块的几何造型组成大面积的几何构成，这使得空间有了延伸感与扩张力。沙发背景墙和酒柜都使用了统一的板材，又在板材中搭配了皮质的材料。

卧室与书房以及房间门口处的更衣间，连成了一个比较私人的大空间。在材料的应用上，依然延续了客厅的软包和板材相结合的设计手法。

This project starts from the micro elements, with no fake luxury, no high profile, no pseudo aristocrats. It sticks to some style which can not be defined with any style, thus it is called micro design series.

For the color design, it focuses on high grade grey colors, with auxiliary ochre and coffee colors, avoiding multiple chromatic aberrations. The materials basically center on warm cloth and boards. The color of the living room is divided into three blocks, with grey equaling ochre color, and white as transition on the ceiling. The format makes use of concise micro design approach, displaying the contrast and variation between big and small, many and few.

The TV wall makes use of a quantity of unified materials, soft rolls. Although this is some common materials, under different lighting, with small blocks of geological formats, large area of geological composition is created, creating some expansive feel and tension inside the space. The sofa background wall and the wine board apply the same material, accompanied with some leather materials.

The bedroom, the study and the cloak room at the entrance of the room create some comparative private grand space. As for the application of materials, it also continues the design approach with the combination of soft rolls inside the living room and boards.

和弦悠扬 Melodious Chords

设计公司：福州华悦空间艺术设计机构
设 计 师：胡建国
项目面积：300 m²
项目地点：福建省福州市
主要材料：仿古砖、大理石、壁纸、玻璃、实木花格、手工银箔

Design Company: Fuzhou Huayue Space Art Design Institution
Designer: Hu Jianguo
Project Area: 300 m²
Project Location: Fuzhou in Fujian Province
Major Materials: Antique Brick, Marble, Wallpaper, Glass, Solid Wood Lattice, Handmade Silver Foil

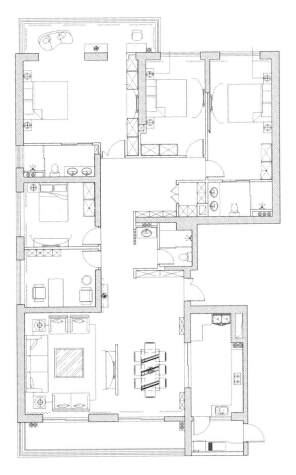

本案是面积约为 300 m² 的单层住宅空间，使用面积较大。设计师在充分满足居室主人对使用功能的要求的前提下，与业主进行沟通，将本案的装饰风格定位为后现代新欧式混搭风格，目的是突破以往的思维模式，做一些大胆的尝试。

在主线条为简欧造型的思路下，设计师采用了带有强烈现代气息的材质来演绎，如不锈钢与皮革的搭配、镜面玻璃与欧式卷草纹样的搭配、木作镂空隔断与手工银箔的对比，无不体现出强烈的现代工业气息。

在色彩搭配上，设计师强调营造出简洁明快的氛围。线条的造型硬朗而不失柔美，装饰元素既延续了欧式风格特有的柔美与和谐，又不失新颖、独特的气质，就像是跌宕起伏的音符演绎出一曲和谐的乐章。

This is a single storey residence with about 300 m², with a large usable area. Upon meeting with the property owner's requirements towards practical functions, the designer communicated with the property owner and set the decoration style of this project to be post-modern neo-European mix and match style with the purpose of breaking through former thinking model and making some courageous attempts.

The principal line is concise European format. The designer makes use of materials with intensive modern atmosphere, such as collocation of stainless steel and leather, mirror glass and European grass pattern, contrast of wooden hollow-out partition and handmade silver foil. All display strong modern industrial atmosphere.

For the collocation of colors, the designer emphasizes on creating concise and brisk atmosphere. The format of lines is kind of tough but soft as well. The decorative elements not only inherit the elegance and harmony exclusive to European style but also feature novel and peculiar temperament, just like musical notes of ups and downs which perform some harmonious musical chapters.

白色的暖意 White Warmth

设计单位：上海唐玛空间设计有限公司	Design Company: Shanghai Tangma Space Design Co., Ltd.
设 计 师：施旭东	Designer: Shi Xudong
项目地点：福建省福州市	Project Location: Fuzhou in Fujian Province
项目面积：130 m²	Project Area: 130
主要材料：复合石材、镜面不锈钢、定制木丝面、皮革硬包	Major Materials: Composite Stone, Mirror Surface Stainless Steel, Custom Make Wood Wool Surface, Leather Hard Roll
摄 影 师：周跃东	Photographer: Zhou Yuedong

本项目是位于闽江边上的高档住宅。设计师以现代的设计手法营造出了典雅、简洁的室内氛围。

推开大门，古典风格的艺术品在灯光下静静地注视着美丽的江景，为这个时尚的空间注入了暖暖的情意。暖色的原木地板、精心挑选的白色理石的天然纹理如江水的涟漪，暗合主人的亲水情结。黑色的皮草和白色花艺、极具现代感的白色装置艺术品，其默契的搭配使得室内氛围和谐而雅致。

整个空间的设计充分将人们的视线打开，运用借景的手法把户外江景引入室内。其色彩、造型均以简洁的手法来呈现，别具写意与悠闲，留给人遐思的空间。

This is a high standard residence on the riverside of Minjiang River. The designer makes use of modern design approach to create some elegant and concise interior atmosphere.

Upon opening the door, classical style artistic objects echoes with the nice river views, instilling some warm sensations for the fashionable space. Warm color log wood floor and carefully selected natural texture are like the ripples of river, corresponding with the complex of the property owner with intimacy with water. The combination of black fur, white floriculture and white decorative artistic objects of modern fell makes the interior atmosphere harmonious and graceful.

The design of the whole space fully opens people's sight and introduces outside riverscape inside with the approach of borrowed scenery. All the colors and formats are all displayed with concise approaches, with poetic and leisurely style, leaving some imagination room for people.

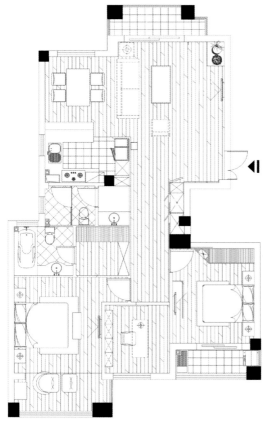

黑与白的邂逅
The Encountering of Black and White

设 计 师：黄耀国
项目地点：福建省福州市
项目面积：80 m²
主要材料：仿古砖、水曲柳面板、明镜、壁纸、水泥漆

Designer: Huang Yaoguo
Project Location: Fuzhou in Fujian Province
Project Area: 80 m²
Major Materials: Antique Brick, Ashtree Panel, Mirror, Wallpaper, Cement Paint

黑、白两色被称为永恒的经典色，不老的传说。在本案中，黑、白、灰三色在空间中有着诸多的表现，就像音乐的三重奏一样，将整个空间打造成轻、重、缓三个不同的音阶。每一个音阶都有不俗的表现，演绎出时尚又略带有古典风情的经典空间。

本案面积较小，在面积较小的居室中设计出舒适且大气的空间就需要设计师的精心构思了。客厅中，大面积的亮色调为空间的主旋律，只有在沙发的边沿、相框的四周镶嵌了黑色的沿边，突显出后现代风格的装饰感。电视墙上的凹凸软包非常富有层次感，也表现出居室空间的高雅品质。沙发背景墙上有一组设计师精心挑选的黑白背景的装饰画，在顶棚筒灯的照射下成为墙面的主角，与色彩纯净、淡雅的墙面形成对比。

Black and white colors are called eternal classical colors and ageless legend. For this project, black, white and grey colors all have multiple presentations in the space, just like trio in music which bring about gentle, heavy and slow – three different musical scales in the space.

This project has a small area. It needs the designer's meticulous conception to create comfortable and magnificent space inside this small area residence. Inside the living room, large area bright color tone is the main theme of the space. There is only black lining which decorate the periphery of sofa and photo frames which highlight the decoration of post-modern style. The concave-convex soft rolls of TV background wall are rich in layers, displaying the elegant quality of the residential space. There is a decorative painting with black and white as the background which the designer carefully selected on the background wall of the sofa, which becomes the leading role of the wall illuminated by the tube lights on the ceiling, which creates some contrast with the wall which has pure and elegant colors.

赏·秋 Appreciation·Autumn

设计单位：珠海空间印象建筑装饰设计有限公司
主案设计师：库瑞辅
软装设计师：潘敏意
项目面积：103 m²
项目地点：广东省珠海市

Design Company: Zhuhai Space Impression Architectural Decorative Design Co., Ltd.
Leading Designer: Ku Ruifu
Soft Decoration Designer: Pan Minyi
Project Area: 103 m²
Project Location: Zhuhai in Guangdong Province

客厅的沙发充满现代感又不失古典意蕴，皮革所呈现的光泽散发出一种高品质的生活格调，古今元素的融合体现出主人深厚的文化底蕴。每件家具和配饰的摆设始终围绕主题，窗帘、地毯的色彩与图案展现出了大气的华美之感。灯饰充满层次感的设计更符合主人的个性，餐桌上精雕细琢的欧式雕花反映出居室主人对品质生活的追求。

主人房中，皮革椅背的大床沉稳之中没有丝毫浮躁感，金属摆件的冷峻使得整个空间沉静下来，显露出一份返璞归真的大气。灯饰的暖橙色调，使卧室多了一份静谧之感。

The sofa inside the living room is full of modern feel, but not lack in classical connotations. The luster of the leather sends out some high quality life style. The integration of ancient and modern elements displays the profound cultural connotations of the property owner. The furnishing and decoration of every furniture and ornament always center on the theme. The color and graphics of curtain and carpet displays the magnificent beauty of the space. The lighting accessories rich in layers are in accordance with the characteristics of the property owner. The refined European style carving on the dining table reflects the property owner's pursuit for quality life.

Inside the master room, the sedate grand bed with leather back does not have any sense of fickleness. The rigidness of the metal accessories makes the whole space quiet down, presenting some magnificence of returning to the original simplicity. The warm orange color tone of the lighting accessories adds some tranquility for the bedroom.

设计单位：珠海空间印象建筑装饰设计有限公司
主案设计师：库瑞辅
软装设计师：潘敏意
项目面积：89 m²
项目地点：广东省珠海市

Design Company: Zhuhai Space Impression Architectural Decorative Design Co., Ltd.
Leading Designer: Ku Ruifu
Soft Decoration Designer: Pan Minyi
Project Area: 89 m²
Project Location: Zhuhai in Guangdong Province

绿色是能表现出大自然勃勃生机的色彩，而且中性纯度的绿色还能起到放大空间的作用。燥热的夏日，在经过一天繁忙的工作后回到家中，祥和的绿色能使人心境平和，再加上绿色植物的装点，整个空间更加生机勃勃，这就是幸福生活的开始。

本案中，客厅墙面上应用了采用自然元素制造的壁纸，比一般的壁纸更加透气。原木雕刻的圆盘在色调相对较深的壁纸上显得活泼、灵动。餐桌上花艺的高度、沙发高度、白色茶几上的摆设、墙上的挂画、主人房的床品和地毯、儿童房的布艺和摆件，甚至各个窗户的布艺颜色，这一切都布置得恰当而妥帖。

客厅的米色沙发，绿、白搭配的布艺地毯，陶瓷质地的花瓶摆件，看似简单的搭配，将陶瓷、布艺、木饰面等紧密结合起来，散发出自己独有的气质。卧室以米绿色为主调，为空间增添了灵动、活泼之感，温馨与舒适之中又体现出生活的情趣。这也是一种时尚考究的生活态度，也是一种永不停歇的居住哲学。

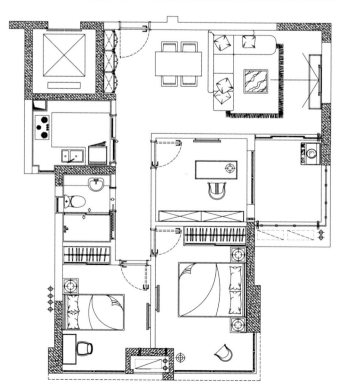

Green is the color that can present the vigor and exuberance of grand nature, and green color with neutral purity can visually expand the space. Suring the hot summer, after a whole day's toil, when you get home, you can get complete relaxation within this kind greenness. With the decoration of green plants, the whole space appears much more vigorous. This is the beginning of happy life.

For this case, the wall of the living room applies wallpaper made of natural elements, which ventilates better than ordinary wallpaper. The round plate carved with log wood appears vivid and dynamic in the comparatively darker wallpaper. All decorations and ornaments are appropriate and proper, such as height of floriculture on the dining table, height of the sofa, the decoration of the white tea table, painting on the wall, bedding accessories and carpet of the master bedroom, fabrics and ornaments in the children's room, even the color of fabrics for every window.

Inside the living room, you can find the beige sofa, green and white fabrics carpet, vase of ceramics texture… which seems simple, but closely combines elements such as ceramic tile, fabrics, wood veneer, etc., displaying its peculiar temperament. Green is the tone color for bedroom, creating some dynamic and vivid sensations for the space. Among this warmth and comfort, there is the life interest. This is some fashionable and exquisite life attitude. This is also some dwelling philosophy with no endings.

设计单位：珠海空间印象建筑装饰设计有限公司
主案设计师：霍承显
参与设计师：王刚
项目面积：345 m²
项目地点：广东省珠海市
主要材料：大理石、镜钢、皮革、橡木、烤漆板

Design Company: Zhuhai Space Impression Architectural Decorative Design Co., Ltd.
Leading Designer: Huo Chengxian
Associate Designer: Wang Gang
Project Area: 345 m²
Project Location: Zhuhai in Guangdong Province
Major Materials: Marble, Mirror Steel, Leather, Oak Wood, Baking Finish Board

设计师在与居室主人进行沟通后，就想利用房子的复式结构打造"空间之中的第二种生活"，二层仍是一家人共同的生活空间，而一层将被设计师打造成具有纯粹生活状态的空间。一层空间中的浪漫氛围扑面而来。顶棚上的木条天花、游走的云彩为室内营造出柔美、温馨的氛围。厨房是开放式的，是一处温馨的就餐环境。夫妻俩的共同爱好是音乐、电影和书籍，设计师特意在沙发旁设置了一个小巧、造型简约的书架，可以在上边放置图书、光碟等。

米黄色调始终贯穿于整个空间中。卧室中，一张简约的大床让人产生躺在上面休憩的冲动。一觉醒来，睁眼便是落地窗外的美景。黄色隔断一侧的浴室，完全对外敞开。何必要遮掩呢？不限定身份、角色的生活状态正是如此的随心所欲。

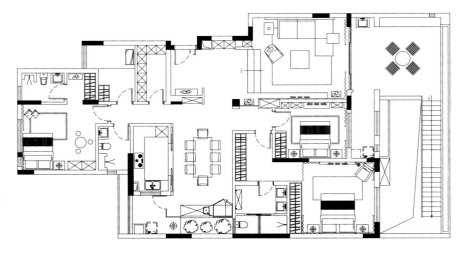

After negotiation with the property owner, the designer wants to create "some other life within the space" through the duplex structure of the house. The second floor is still the common living space for the family, and the first floor is designed into a space with pure life conditions. The first floor is overflowing with romantic atmosphere. The wood ceiling and migrating clouds on the ceiling create soft and warm atmosphere for the interior space. The kitchen has open style which is a cozy dining space. This couple has a lot in common, such as likes for music, films and books. The designer specifically arranged a tiny book shelf with concise format beside the sofa, where people can store books, disks, etc.

Beige color tone runs through the whole space. Inside the bedroom, this huge simple bed makes people want to sleep on it and have a complete rest. Once you wake up, what you see are all terrific views outside the French window. The bathroom on one side of the yellow partition is utterly open. Why need to hide? The life conditions with no strict status or characters are so much at one's own will.

跳跃的精灵
Dancing Angles

设计公司：广州共生形态工程设计有限公司
www.cocopro.cn
项目地点：广东省佛山市

Design Company: Guangzhou C&C Design Co., Ltd.
www.cocopro.cn
Project Location: Foshan in Guangdong Province

后现代新古典

森林里随风扬起的花瓣落在墙上，形成了色彩丰富的挂画。玫瑰造型的茶具、俊丽的黑色羚羊书架，还有喇叭造型的落地灯聚集在一起，仿佛像是森林精灵们的大聚会。

餐厅中神秘魔术师的帽子灯、不规则的错拼装饰画、彩色菱形格的餐垫，再点上蜡烛，温馨浪漫的晚餐原来在家里也能实现。

书房中，森林植物图案的壁纸，帅气的小老虎书档，机灵的兔子台灯，让您在工作之余，也能够体会到来自于大自然的乐趣。

转身步入主卧，飘窗台上特别定制了梳妆柜，方便女主人的生活。宽大的飘窗台可供主人休闲时的看书、听音乐，在此享受每一个日出与日暮，是一种难得的享受。

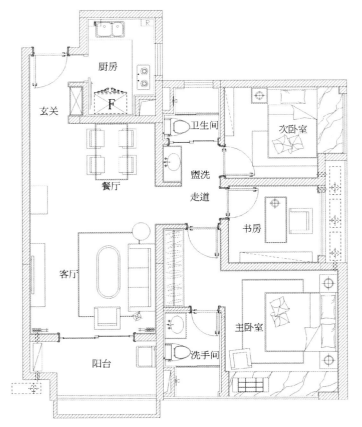

Petals in the forest dancing in the wind fall on the wall, forming a painting with abundant colors. The tea set of rose format, nice black antelope book shelf together with floor lamps of horn modeling gather here, which is like a grand get-together for spirits in the forest.

Inside the dining hall, there are hat lamp of mysterious magician, irregular decorative painting, table mats of colorful diamond-type lattice. While the candles are lit, people can also enjoy romantic dinner inside the house.

Inside the study there are wallpaper of plants graphics from the forest, handsome bookends of little tiger pattern, dynamic rabbit lamps. When you are free from work, you can also experience the fun of the grand nature.

Inside the master bedroom, there are custom-made dresser on the windowsill, for the convenience of the hostess. The grand windowsill can allow the host to read books and enjoy the music during the leisure time. What kind of enjoyment to enjoy every sunrise and every sunset here.

光大锦绣山河样板房
Show Flat of Ever-bright Splendid Landscape Property

设 计 师：王五平
项目地点：广东省东莞市
项目面积：170 m²
主要材料：多乐士乳胶漆、红橡木油白、灰境、全抛釉砖、壁纸

Designer: Wang Wuping
Project Location: Dongguan in Guangdong Province
Project Area: 170 m²
Major Materials: Dulux Emulsion Paint, Grey Mirror, Glazed Brick, Wallpaper

后现代新古典

摩登样板间 II

该项目坐落于风景秀丽、景色怡人的东莞市松山湖行政商务区的核心地段,东临碧波万顷的松山湖,西依翠绿的生态园林,可谓居住地带的天然氧吧。

在平面的处理上,本案设计的最大亮点就是设计师打造了一个超大面积的主卧,里面有卧室、衣帽间、大洗手间、书房、休闲塌塌米等,是夫妇二人休憩时间的最佳去处。

本案的户型是跃层复式,餐厅比客厅的层高要高一些,相对来说比较独立。餐厅设计是本案的一个特色,立面的造型延伸到顶棚,两侧的酒柜造型简洁。客厅背景墙运用了黑色的大理石砖,设计师将电视放置进大理石打造的凹槽里,功放机、机顶盒放置在右边的暗柜里,电视墙的造型简洁、干净而大气。在色彩上,本案大胆地运用黑色,与白色形成鲜明的对比,营造出一种简约、干净的效果。

This project is located in the core area of the administrative business area of Songshan Lake district of Dongguan which is a picturesque area, with Songshan Lake of immense green waves on the east side, and a verdant ecological garden on the west, which is quite a natural air anion bar as a residential area.

For the arrangement of plan, the biggest bright spot for this design is the huge size master bedroom where there are bedroom, cloakroom, grand washroom, study, matting, etc. which is a perfect space for the couple to spend their leisure time.

This case has duplex flat type. The storey height of the dining hall is higher than the living room, thus comparatively independent. The design of the dining hall is one characteristic of the project. The format of the façade extends towards the ceiling, and the wine cabinets on both sides have concise modeling. The living room's background wall applies black marble bricks. The designer set the TV inside the groove made of marble. The audio-power amplifier and the STB were placed in the cabinet on the right side. The modeling of TV background wall is concise, clear and magnificent. For colors, this project courageously made use of black to form bright contrast with white color, producing some concise and tidy effects.

设计单位：广州共生形态工程设计有限公司
WWW.COCOPRO.CN
项目地点：广东省中山市

Design Company:Guangzhou C&C Design Co., Ltd.
WWW.COCOPRO.CN
Project Location: Zhongshan in Guangdong Province

这是一套标准样板房，生活在其中的主人是一位精明能干的时尚达人，工作之余他注重生活的情趣，热爱歌剧与音乐。

说到歌剧《卡门》，总让人联想到爱情与冒险。在本案中，设计师希望赋予原本平淡无趣的室内空间以生命和个性。色彩上以咖啡色、米白色调为主。设计师比较注重材料质感的对比，通过皮革、马毛、壁纸和不锈钢等材料的强烈对比，体现一种精致的奢华感和男性化的生活方式与氛围。结合居室主人对歌剧的理解，设计师希望在此生活的人能够享受到一份极致的生活情趣。

This is a standard sample house. The master living inside is an intelligent and hard-working fashion insider. After work he focuses on the interests of life and has great passion for opera and music.

The opera Carmen would often make people think of love and adventure. For this case, the designer aspires to give the original insipid and joyless interior space life and individuality. For colors, the designer focuses on coffee color and beige. The designer pays a lot of attention to the comparison of materials' texture. Through the intensive contrast of materials such as leather, horsehair, wallpaper and stainless steel, there is the presentation of exquisite luxury and masculine lifestyle and atmosphere. Combined with the understanding of the master towards opera, the designer hopes that people living here can enjoy some extreme life interests.

后现代新古典

设计单位：简艺东方设计机构
设 计 师：孙长健　林元娜
项目面积：256 m²
主要材料：进口金刚板、枫木地板、进口壁纸、爵士白大理石、木纹石、布艺仿皮软包、仿古通体砖
摄 影 师：吴永长

Design Company: Simple Art Oriental Design Institution
Designer: Sun Changjian, Lin Yuan'na
Project Area: 256 m²
Major Materials: Imported Laminate Flooring, Maple Wood Flooring, Imported Wallpaper, Jazz White Marble, Wood Pattern Stone, Cloth Imitation Leather Soft Roll, Antique Brick
Photographer: Wu Yongchang

后现代新古典

这是一套四层的叠拼别墅。居室主人喜欢色彩明亮、风格简约、又略带一丝淡雅的奢华与尊贵感的生活空间。因此，设计师将总体的色调定位为明快的浅黄色，局部采用大块面深咖色及黑胡桃木色为协调色，在原本平淡的浅黄色空间中，增添了一些稳重感，使整体空间的层次更加分明，塑造出节奏明快的氛围。

在软装的设计上也进行了特别的设计，客厅中的地毯将总体色调与视觉效果融洽地结合起来。抱枕、装饰挂画、花艺、装饰摆件等恰到好处的陈设设计，使整体空间散发出淡淡的儒雅气质，将艺术性与实用性完美地结合起来。由于客厅空间的原始结构偏长，整体空间的视觉感不佳。因此，我们在设计上特意做了一个抬高的休闲平台，既丰富了整体空间的层次，同时又多出了一个使用空间。日常生活中，这里是小孩玩耍的平台，大人也可以在此席地而坐，或者读书品茗，生活变得更加惬意。

餐厅与厨房空间应用了全通透的玻璃进行分隔，既满足了使用功能，也能使得空间更加通透，色调上与客厅空间相协调，不规则的木纹石成了餐厅中的视觉焦点。

主卧室原本的结构不是太理想，于是我们将书房与主卧的位置做了调换，将楼梯处的小起居室融入到了主卧套间中，将主卧室设计成了一个集大主卫、书房、更衣间、休息室于一体的大套间。设计上既注重了空间与空间之间的视觉通透感，又保证了光线能够到达每一个角落，使得整体空间更加开阔。

This is a 4-storey townhouse. The property owner enjoys living space with bright colors, concise style and a trace of elegant luxury and aristocratic feelings. The designer set the whole tone color to be brisk light yellow colors and applied chunks of dark coffee color and black walnut color as the coordinating colors. This adds some steady feeling to the original insipid light yellow space to make the whole space with distinct layers, thus creating atmosphere with vivid rhythms.

The soft decoration also applies some specific design to harmoniously associate the whole color tone with the visual effect. The to-the-point layout design of pillows, decorative paintings, floriculture and decorative objects makes the whole space send out some elegant temperament, which perfectly

combines artistic quality and practical quality. Due to the original long structure of the living space, the whole space does not have fine visual quality. Thus the designer specially makes a raised leisure platform which not only enriches the layers of the whole space, but also creates some other practical space. For daily life, this is a platform where children can play, and adults can also sit on the ground, reading books or enjoying tea. Life is made much more colorful.

The dining hall and the kitchen use wholly transparent glass as the partition which not only carries out practical functions, but also makes the space appear much more transparent. The color tone is in accordance with the living room. And the irregular wood grain stone becomes the visual focus of the dining hall.

The original structure of the master bedroom is not perfect. Thus the designer makes some changes towards the location of study and master bedroom, and integrates the small living room around the staircase into the master bedroom suite, thus changing the master bedroom into a grand suite with large master bath room, study, dressing room and lounge. The design not only focuses on the visual transparency among spaces, but also guarantees that light can get to every corner, thus making the whole space appearing broader.

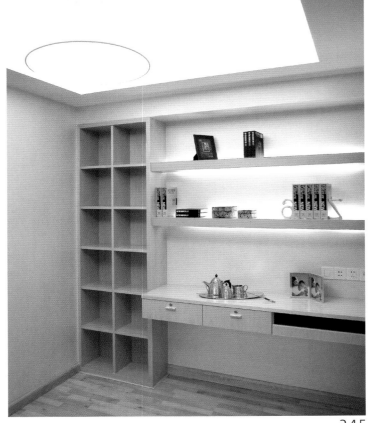

设计单位:简艺东方设计机构
设 计 师:孙长健 林元娜
项目面积:120 m²

Design Company: Simple Art Oriental Design Institution
Designer: Sun Changjian, Lin Yuan'na
Project Area: 120 m²

这是一套沿江的住宅，设计师考虑到江景风光以及行船声音等环境因素，将书房与主卧做了位置上的互换，既减少了原有过道空间的浪费，又使得客厅与书房外的景致融为一体，同时保持了卧室的宁静感。

客厅与餐厅之间有一根很大的承重梁，因此这大弧形的吊顶造型既隐藏了笨重的梁体，同时又将入门的鞋柜包裹其中，保证了整个空间的纯净与整体性。餐厅的弧形吊顶起到了很好的缓冲效果，在保证餐厅高度的同时，又保证了视觉上的纯净感。在单纯的色彩与造型中，整体空间配上梯田式样的造型，在保持整体统一的同时，又带来了一点活跃的氛围，为居住者带来了一份纯净的心情。

This is a residence along the river. Considering environmental elements such as river-scape views and noise of sailing boats, the designer makes some interchange towards the location of study and master bedroom, which not only decreases the waster towards the original corridor space, but also integrates the living room with the views outside the study, while at the same time, the tranquility of the bedroom is maintained.

There is a huge spandrel girder between the living room and the dining hall. This arch ceiling not only hides the heavy girder, but also encloses the shoe cabinet at the entrance inside, securing the purity and integrity of the whole space. The arch suspended ceiling achieves some perfect buffering effect. While maintaining the height of the dining hall, it assures the visual purity. Within the simple colors and formats, accompanied with terrace pattern modeling, the whole space displays integrity and consistency, while bringing some vibrant atmosphere, creating some pure moods for the people living here.

长宁复邦住宅
Fubang Residence in Changning District

设计师：赖仕锦	Designer: Lai Shijin
项目地点：上海市	Project Location: Shanghai
项目面积：150 m²	Project Area: 150 m²
主要材料：饰面板、黑镜、木地板、壁纸	Major Materials: Veneer, Black Mirror, Wood Flooring, Wallpaper
撰　　文：赖仕锦	Composer: Lai Shijin

后现代新古典

这是一套附带阁楼的住宅，可惜阁楼分在两边，还没有楼梯。因此，在不大的空间中加两个楼梯是很棘手的问题。设计师巧妙地通过空间转换，在一个合适的位置设置了主楼梯，另一个不常用的楼梯则采用伸缩式的。这样一调整，空间就变得丰富起来。

在墙面的装饰上，设计师采取了简洁化的处理方式，并利用大面积的整体材质将储物空间隐藏起来，最后呈现出来的效果大方而不失整体感。在色调上，设计师将亮色调与暗色调做了强烈的对比，再加上褐色调的调和，空间的层次以及表情显得更加丰富。

This is a residence with an attic which is set on both sides, and there are no stairs originally. Thus it would be not easy for the designer to add two stairs in the small space. The designer ingeniously made space switching and set the main stair in the appropriate location. And the other not frequently-used staircase is made retractable. Through such arrangement, the space is made colorful.

For wall decoration, the designer applied concise manipulation approach and hid the storage space with large area of materials. The final effect is generous and does not lack in integrity. For color tone, the designer made some sharp contrast between bright color tone and dark color tone. Accompanied with the reconciliation of brown color, the layers and expression of the space appear richer and more abundant.

名都城住宅 Mingducheng Residence

设 计 师：赖仕锦	Designer: Lai Shijin
项目面积：140 m²	Project Area: 140 m²
项目地点：上海市	Project Location: Shanghai
主要材料：饰面板、黑镜、木地板、软包	Major Materials: Veneer, Black Mirror, Wood Flooring, Soft Roll
撰　　文：赖仕锦	Composer: Lai Shijin

后现代新古典

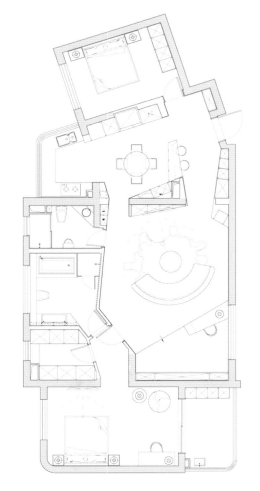

本案的业主为一位金融界的成功人士，充满男性的阳刚之气，并有浪漫的情怀。根据居室主人的要求，设计师要从客户自身的条件及性格出发，塑造出充满个性而又浪漫、神秘的空间氛围。

本案原有的空间格局零散而又封闭，客厅的光线也比较昏暗，接触自然光不多，因此，设计师最大的任务便是改变这种现状，将自然光线引入室内。在征得业主同意的情况下，施工者打通了部分非承重墙面，巧妙地利用每个墙垛连接起一个比较规整的开放式空间，格局上显得生动而又活泼。同时，又利用了南、北方向照射进来的部分自然光线，通过色调、材质的深浅对比，以及对照明效果的控制，营造出了理想中的充满个性而又男性气质十足的魅力空间。

The property owner is a successful finance guy full of masculinity and romantic sensations. According to the requirements of the property owner, the designer needs to start from the owner's own conditions and characteristics, thus creating romantic and mysterious space atmosphere full of individualities.

The original space structure is scattered and confined, and the lights of living room are kind of dim, without much natural light pouring inside. With the agreement of the property owner, the constructor broke through part of nonbearing wall. Piers are skillfully connected to form an open space in order which makes the layout appear lively and vivid. At the same time, the designer makes use of the light coming inside from south and north, dark and light contrast of color tones and materials and control of illumination effect, thus producing a space full of charms with characteristics and masculinities.

清水湾住宅
Clear Water Bay Residence

设 计 师：赖仕锦、吴义云
项目面积：120 m²
项目地点：上海市
主要材料：染色饰面板、黑镜、地砖、黑色不锈钢
撰　　文：赖仕锦

Designer: Lai Shijin, Wu Yiyun
Project Area: 120 m²
Project Location: Shanghai
Major Materials: Dyed Veneer, Black Mirror, Floor Tile, Black Stainless Steel
Composer: Lai Shijin

原来的空间中，空间结构比较呆板、单调。本案的主人要求室内设计能够改变这种比较压抑的现状。在经过与业主的充分沟通以及协调后，一个全新的空间构想便随之诞生了。

墙面大块面不规则的转折、顶棚富有动感的艺术造型、蓝色与白色相间的方块墙面等，都使得这个空间在视觉上给人以刺激、新潮的感觉，比较符合居室主人求新、求变的要求。

在装饰材料上，公共空间的地面使用了浅色的木纹石地砖，卧室使用了实木地板，墙面以乳胶漆以及染色饰面板为主，顶棚使用了黑色的不锈钢收边。电视背景墙的色块造型错落有致，原来相对呆板的矩形空间变得更加丰富多彩。居室主人对改造后的空间非常满意。通过设计师巧妙的表现手法，原本的不如意变成了惊喜。

For the original space, the structure is monotonous and rigid. The master hopes that the interior design can change the depressing current situation. A brand-new space comes into being after sufficient communication and negotiation with the property owner.

Chunks of irregular transitions on the wall, dynamic artistic modeling on the ceiling and diamond square wall surface all make this space stimulating and trendy, in accordance with owner's demands for novelty.

As for decorative materials, the floor of the public space applies light color wood grain floor tile. The bedroom applies solid wood flooring. The wall mainly uses emulsion paint and dyes veneer, and the ceiling is lined with black stainless steel. The color modeling of the TV background wall is well-proportioned, thus the original rigid rectangular space is made much more colorful. The property owner is very satisfied with the transformed space. Through ingenious approach of the designer, the original commonness becomes surprise.

万科缤纷夏日样板房
Show Flat of Vanke's Wonderful Summer Property

设计单位：广州共生形态工程设计有限公司
WWW.COCOPRO.CN
项目地点：广东省佛山市

Design Company: Guangzhou C&C Design Co., Ltd.
WWW.COCOPRO.CN
Project Location: Foshan in Guangdong Province

活力四射的夏日都市中,简约风格的住宅也要激情洋溢。

客厅中,错拼的彩色长条装饰画搭配白色简约风格的沙发以及波浪形、彩色条纹的抱枕,就连蒲公英都忍不住要随风起舞。

餐厅中,嫩嫩的草绿色调的桌面摆设、雏鸟形的餐巾扣、白色简约的树形烛台完美地搭配在一起,营造出明亮夏日里一抹醉人的惬意。精致明亮的厨房,彩色艳丽的精巧厨具,制作出可口的美食来诱惑你我的感官与味蕾。

儿童房中有翠绿色的窗帘、小巧的木马和缤纷的泡泡床品,活泼的色彩表现了孩子天真可爱的性情。

Within this dynamic summer city, residence of concise style should be overflowing with enthusiasm as well.

Inside the living room, the colorful strip decorative painting is accompanied with sofa of concise style and wave-like throw pillow of colorful strips. Even dandelion can't wait to dance in the wind.

Inside the dining room, the table accessories of grass green color, fledgling bird-like napkin button and white tree-like candlestick are perfectly combined together to create some pleasing comfort in the bright summer. Exquisite and bright kitchen and colorful kitchenware produce delicious food to seduce our sense organs and taste bud.

Inside the children's room, there are emerald curtain, tiny wooden horse and colorful bedding accessories. The active colors display the naivety disposition of children.

五矿·哈施塔特别墅样板房
Wukuang·Hallstatt Villa Show Flat

设计公司：深圳张起铭室内设计有限公司
设 计 师：张起铭
项目面积：500 m²
主要材料：橡木饰面，壁纸，大理石
摄 影 师：钱祥

Design Company: Shenzhen Zhang Qiming Interior Design Co., Ltd.
Designer: Zhang Qiming
Project Area: 500 m²
Major Materials: Oak Wood Veneer, Wallpaper, Marble
Photographer: Qian Xiang

后现代新古典

本案位于广东省东南部的惠州，外倚绵延群山，尽享湖光山色，这里有一处近17万平方米的天然湖泊。项目以独有的东方精神演绎了奥地利特有的湖光山色，其被誉为"来自仙境的明信片"。这个别墅区中，教堂、广场、街区、酒店等生活以及休闲设施一应俱全，其优美的景色像极了一幅风景画，演绎着奥地利哈施塔特的浓郁风情。

本套别墅样板房的室内设计有着绅士的气质，室内软装饰所传达出来的高雅品位一改传统意义上古板、单调的绅士形象，设计师将舒适且随性的室内氛围营造得恰到好处，为居室主人打造了一处安心且愉悦的私人生活空间。

This project is located in Huizhou south-east of Guangdong Province, surrounded by continuous mountains, lake views and mountain landscape. There is a natural lake which is about 170,000 square meters. With its peculiar oriental charms, this project presents the landscape exclusive to Austria which is crowned as "a postcard from fairyland." In this villa area, there are all kinds of life and leisure facilities, such as church, square, block, hotel, etc. The nice view is like a landscape painting, displaying the intense charms of Hallstatt of Austria.

The interior design of the villa show flat possesses temperament of a gentleman. And the elegant taste of interior soft decoration utterly changed the old-fashioned and monotonous image of gentleman in traditional meaning. The designer appropriately created some cozy and casual interior atmosphere, modeling some pleasing and comforting private living space.

后现代装饰主义样板间
Show Flat of Post-Modern Decorationism

设计单位：深圳市艺鼎装饰设计有限公司
设 计 师：王锟
项目面积：108 m²
项目地点：吉林省长春市

Design Company: Shenzhen Yiding Decoration and Design Co., Ltd.
Designer: Wang Kun
Project Area: 108 m²
Project Location: Changchun in Jilin Province

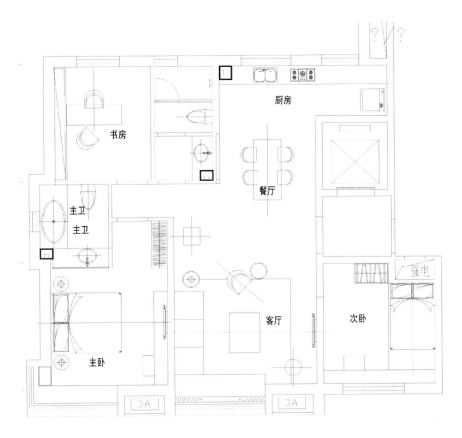

后现代新古典

本案户型规整，原始的平面布局比较合理。因此，设计师的重点在于营造出与主人的品位、兴趣、爱好相协调的居室氛围。在与业主沟通协调后，设计师将之定位为现代家居风格。

客厅以沉稳的褐色调为主。沙发背景墙上的皮质软包与灰镜相互协调，白色的布艺沙发、造型简约的钢化茶几为空间增添了一些洒脱的气质。餐厅与客厅空间相连接，延续了客厅空间低调、简约的格调，木色的墙面、造型简约的木质餐桌椅等都显示出居室主人平和、朴实的生活态度。另外，餐厅与厨房联系在一起，以墙面上不同的装饰材质来区分各自的功能区域，显示出设计上的灵活性。

主人喜好读书，书房这个功能空间是必不可少的。在这个空间中，设计师用木色调的色彩为这个空间营造出古色古香的氛围。又利用造型简约而现代的定制书架、书桌来体现现代风格的居室格调。

This project has regular flat type and proper original plane layout. Thus, the focus of the designer is on creating some residential atmosphere in harmony with the taste, interests and likes of the property owner. After negotiation with the owner, the designer set it to be modern residential style.

The tone color of the living room is brown. The leather soft roll on the background wall of the sofa echoes the grey mirror. White fabrics sofa and concise tempering tea table add some free and easy temperament for the space. The dining hall is connected with the living room, continuing the low-key and concise style of the living room. Wood color wall and concise wood dining table and chairs all present the calm and sincere life attitude of the property owner. Other than that, the dining hall is connected with the kitchen. They use different decorative materials on the wall to divide their specific functional areas, displaying the flexibility of design.

The property owner enjoys reading books. Thus, study becomes a necessary space. For this space, the designer uses wood color to produce antique atmosphere. Apart from that, the designer uses concise and modern custom-made book shelf and table to display the residential tone of modern style.

宜家风格样板房
Show Flat of IKEA Style

设计单位：深圳市艺鼎装饰设计有限公司	Design Company: Shenzhen Yiding Decoration and Design Co., Ltd.
设 计 师：王锟	Designer: Wang Kun
项目面积：96 m²	Project Area: 96 m²
项目地点：吉林省长春市	Project Location: Changchun in Jilin Province

本案是一个约96m²的样板间，风格定位上以宜家风格为主。温馨、浪漫、方便、清爽、简洁就是宜家风格的特点，目的就是营造出适宜人们居住的居家环境。

在本案中，设计师将宜家风格的特点发挥到了极致。整体色调以乳白色、木色为主，局部点缀黑色，混搭出清新、雅致的设计格调。客厅中，沙发、地面以及墙面使用了大面积的乳白色，色调整体又统一，给人以清新雅致的感觉。沙发墙面上呈不规则排列的风景画为整洁的墙面增添了一些别样的情愫。电视墙面上使用了质朴而雅致的木饰面板，打破了乳白色的轻盈、飘逸之感，使得整体空间氛围更加稳重、大气。

This is a show flat which is about 96 m², with the style oriented to be IKEA style. The characteristics of IKEA style can be described as warm, romantic, convenient, clear and concise, with the purpose of creating appropriate home environment for people.

For this case, the designer displays IKEA style to the extreme. The whole tone color is focused on milk white and wood colors, dotted with black, creating fresh and elegant design style. The landscape paintings on the wall arranged in some irregular way create some different charms for the clear wall. The TV background wall applies some primitive and elegant wood veneer, breaking through the airy and graceful atmosphere created by milk white color, which make the whole space appear much more sedate and magnificent.

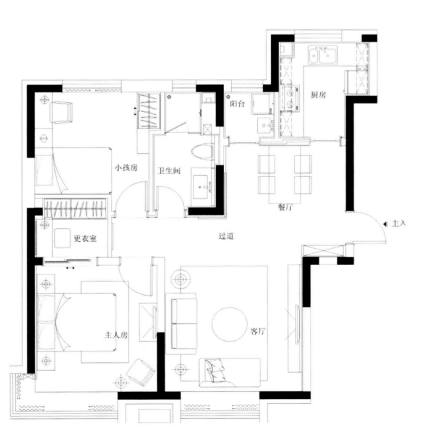

摩登个性之家
Modern and Personal Home

设计单位：J2-DESIGN/厚华装饰设计有限公司
设 计 师：黄炽烽、欧敏华
陈设与选材：区婷婷
开 发 商：广东新中源集团

Design Company: J2-DESIGN/Houhua Consultant Design Co.,Ltd
Designers: Huang Chifeng, Ou Minhua
Layout and Materials Designer: Ou Tingting
Developer: New Zhongyuan Group in Guangdong Province

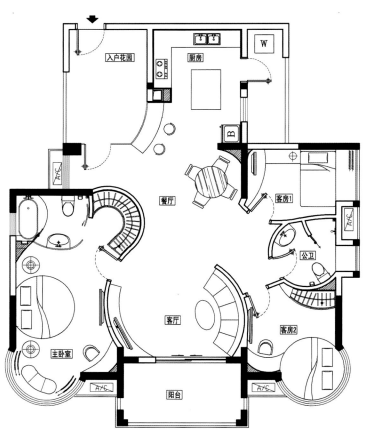

本案为摩登个性风格，设计师在这个空间的设计中打破常规，以曲面空间为主题，将顶棚、墙身、家具融为一体。在视觉上，整个空间赋予人新奇、大胆的体验。

设计材料以长城板、镜面有机片、黑色抛光砖为主，营造出黑与白相映成趣的效果。在平面规划图上设计师将入户花园、厨房、餐厅、客厅连为一体，形成富有动感的曲线空间。主人房与卫生间之间采用通透的钢化玻璃作为隔断，赋予这个空间以独特性，形成了类似酒店客房的氛围。

This project has modern style. For the space design, the designer breaks through the common rules and focuses on curved surfaces, integrating ceiling, wall and the furniture. The space gives people novel and adventurous visual experiences.

The design materials focus on board, mirror surface and black polished tile, creating some effect with black and white colors bringing out the best in each other. For the plane layout, the designer combines hallway garden, kitchen, dining hall and living room, creating some curved space with dynamic feel. The space between the master room and the washroom applies toughened glass as the partition, entrusting the space with peculiar characteristics, forming some atmosphere like hotel guest rooms.

设 计 师：王五平
项目地点：广东省深圳市
项目面积：700 m²
主要材料：灰木纹、大理石砖、乳胶漆、水曲柳油白、皮革、壁纸

Designer: Wang Wuping
Project Location: Shenzhen in Guangdong Province
Project Area: 700 m²
Major Materials: Grey Wood Grain, Marble Brick, Dulux Emulsion Paint, Leather, Wallpaper

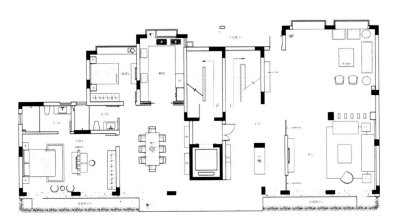

后现代新古典

本案有着超大的视觉平面空间，且南北通透，外边的自然光可以毫无阻碍地流泻进来。在冬日，可以尽情地享受阳光，慵懒地发呆，忘记一切烦杂琐事。在黄昏时分，这里晚风拂面，宜景怡心。

在这样绝佳的空间条件下，设计师需要营造的就是一种洗尽铅华之后的雍容感，空间不仅仅是一处休息的场所，也是一处寄托灵魂的存储室。在空间处理上，各个空间相互渗通，大而不空，力求营造出一个现代、时尚、大气、精致的空间氛围。

This space is transparent in the north and south direction, with grand visual plan. The outside natural light can spread inside with no obstructions. During winter time, people can enjoy sunshine, relax idly and forget about all these trivial things. At dust, people can enjoy the gentle breeze and the terrific landscape.

With this perfect space circumstances, what the designer wants to create is some natural, graceful and poised feeling with all the fancy time over. This space is not only some space for leisure, it is also some storage for the souls. For the space treatment, various spaces infiltrate one another, grand but not vacant. The designer aspires to create some modern, fashionable, magnificent and delicate space atmosphere.

设计单位：北岩设计	Design Company: Beiyan Design
项目地点：江苏省南京市	Project Location: Nanjing in Jiangsu Province
项目面积：150 m²	Project Area: 150 m²
主要材料：壁纸、硅藻泥、硬包、原木家具	Major Materials: Wallpaper, Diatom Ooze, Hard Roll, Log Wood Furniture
摄　　影：金啸文空间摄影	Photographer: Jin Xiaowen Space Photography

本案有着超大的视觉平面空间，且南北通透，外边的自然光可以毫无阻碍地流泻进来。在冬日，可以尽情地享受阳光，慵懒的发呆，忘记一切烦杂琐事。在黄昏时分，这里晚风拂面，宜景怡心。

在这样绝佳的空间条件下，设计师需要营造的就是一种洗尽铅华之后的雍容感，空间不仅仅是一处休息的场所，也是一处寄托灵魂的存储室。在空间处理上，各个空间相互渗通，大而不空，力求营造出一个现代、时尚、大气、精致的空间氛围。

This space is transparent in the north and south direction, with grand visual plan. The outside natural light can spread inside with no obstructions. During winter time, people can enjoy sunshine, relax idly and forget about all these trivial things. At dust, people can enjoy the gentle breeze and the terrific landscape.

With this perfect space circumstances, what the designer wants to create is some natural, graceful and poised feeling with all the fancy time over. This space is not only some space for leisure, it is also some storage for the souls. For the space treatment, various spaces infiltrate one another, grand but not vacant. The designer aspires to create some modern, fashionable, magnificent and delicate space atmosphere.

眷鸟 Juan Niao

设计单位：北岩设计
项目地点：江苏省南京市
项目面积：150 m²
主要材料：壁纸、硅藻泥、硬包、原木家具
摄　　影：金啸文空间摄影

Design Company: Beiyan Design
Project Location: Nanjing in Jiangsu Province
Project Area: 150 m²
Major Materials: Wallpaper, Diatom Ooze, Hard Roll, Log Wood Furniture
Photographer: Jin Xiaowen Space Photography

后现代新古典

遥望天涯路，眷鸟知归途……小鸟眷恋着归途，我们对家也有着深深的眷恋。80后的我们对家有着自己独有的情怀，想拥有一座属于自己的房子，自己布置，自己生活，过一种干净的、自由自在的生活。

在材料应用上，设计师从低碳的角度出发运用了一些新型环保材料，如硅藻泥、硬包、原木家具等，使空间更具质感和层次。在空间布局上稍作调整，原餐厅北阳台被改为厨房，厨房则被改成吧台区，空间结构发生了根本性的改变，空间的利用率也大大增加。

温馨的布艺软装饰品点缀其中，与空间整体相互呼应，旨在营造出一个简约、宁静，又具有自然气息的宜居空间。

Looking afar at the distant in the end of the world, and the homesick bird wants to go back to its nest. Even a little bird longs for home back, we as human beings also have deep passion for home. For people who are born after 1980, we also have some spectacular sensations, striving for a home of our own. Hence we can arrange by ourselves and live inside, having some clear and free life.

For the application of materials, the designer also explores some new environmental materials from low-carbon aspect, such as diatom ooze, hard roll, log wood furniture, etc., making the space full of layers and texture. The designer carries out some tiny adjustments towards the space layout. The north balcony of the original dining hall is changed into a kitchen, and the original kitchen is changed into a bar counter. Thus there are drastic transformation towards the space structure, while improving the use ratio of the space.

The space is dotted with warm cloth soft decoration objects, corresponding with the whole space and creating a concise and quiet space, making the space become some livable space with natural atmosphere.

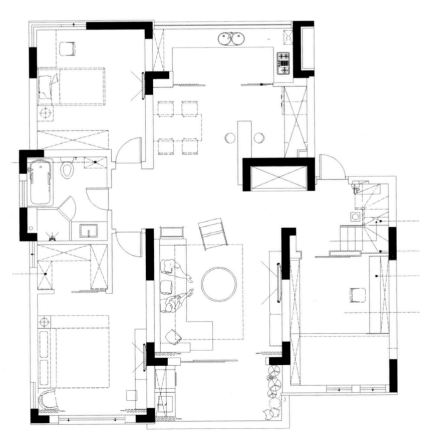

苏州水岸枫情公寓
Riverside Maple Apartment in Suzhou

设计单位：威利斯设计有限公司
项目面积：130 m²
项目地点：江苏省苏州市
主要材料：地板、壁纸、皮纹砖、仿木纹砖、陶瓷锦砖、镜子、不锈钢条

Design Company: Wu Xiaowei/Willis Design Center
Project Area: 130 m²
Project Location: Suzhou in Jiangsu Province
Major Materials: Floor, Wallpaper, Leather Grain Tile, Imitation Wood Grain Tile, Mosaic Tile, Mirror, Stainless Steel Strip

生活是丰富多彩的，家居生活同样如此。当很多人还追求横平竖直、中规中矩的线条时，我们的设计师却另辟蹊径，通过斜线来营造出空间感，点、线、面相结合，给人以强烈的视觉冲击，打造出移步换景的闲适空间。

首先，设计师对房屋的结构做了一些改造，拆除了部分墙体，整个空间看起来更为连贯、开阔，同时改造出了一些特殊空间，进门处除了必要的鞋柜外，沿着墙的转角处还有一个简单的吧台，门厅与餐厅之间的过渡一下子变得生动起来。

客厅的设计极为独特，餐厅和客厅应用了不同材质的地砖和地板，餐厅中部分采用地板色饰面板装饰吊顶和斜切的墙体，给人视觉上的错位感与变形体验，有着强烈的后现代主义情愫。地面也通过咖啡色的皮纹砖、原木色地板等的斜铺突破了传统家居的框架，给人耳目一新的感觉，会客区的层次和空间感更为强烈。

主卧的背景墙两侧采用斜切的如水墨画般的壁纸装饰，辅以不锈钢装饰架，给原本简约的空间带来无限的张力。

Life is colorful, so is home life. When a lot of people are still pursuing straight, vertical and horizontal lines, our designers take another way, creating a space with oblique lines. The combination of spots, lines and surfaces leaves people with intensive visual impact, producing a space of varying views.

First of all, the designer makes some transformation towards the structure of the house. Through removing part of the wall, the whole space was made consistent and broad. While at the same time, the designer created some special space. Aside from the necessary shoe cabinet at the door entrance, there is a simple bar counter at the corner of the wall. The transition between the hallway and the dining hall becomes much more dynamic.

The design of the living room is very peculiar. The dining hall and the living room use floor tiles and floor boards of different materials. Part of ceiling and the wall inside the dining hall applies floor board color veneer as the decoration, leaving people with visual dislocation impression and transformation experiences, with intensive post-modern sensations. The floor is paved with coffee color leather grain tiles and log wood color flooring which break through the traditional framework of home furnishing, leaving people with brand-new impressions. And the sense of layers and spaces of the living room is made more intensive.

Both sides of the background wall inside the bedroom use ink-painting-like wallpaper as the decoration. Accompanied with stainless steel as the decorative frame, the original concise space is supplied with limitless tensions.

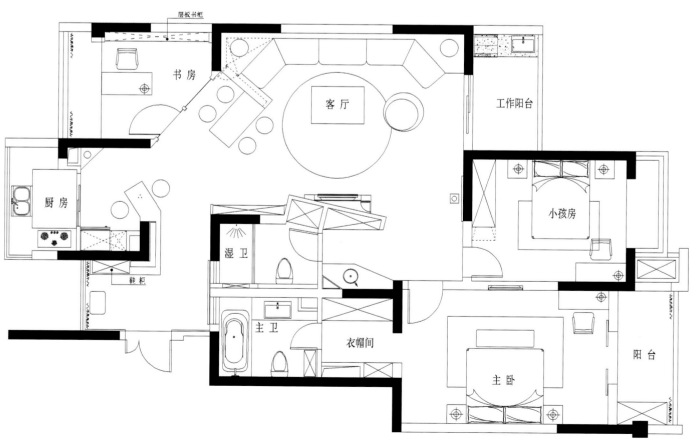

常熟世茂三期样板间
Show Flat of 3rd Phase of Shimao Property in Changshu

设计单位：威利斯设计有限公司
设 计 师：杨旭
项目面积：130 m²
项目地点：江苏省苏州市
主要材料：壁纸、软包、钛合金、银镜、茶镜、仿古砖、地板

Design Company: Willis Design
Designer: Yang Xu
Project Area: 130 m²
Project Location: Suzhou in Jiangsu Province
Major Materials: Wallpaper, Soft Roll, Titanium Alloy, Tawny Mirror, Antique Tile, Flooring

后现代新古典

本案为一套三室两厅的平层公寓，以现代简约风格为主，大量使用原木色饰面板以及软包装饰墙面，赋予空间安逸、祥和之感。此外，还在局部使用钛合金、银镜等质感十足的现代工艺来点缀，空间又变得灵动、跳跃，耐人寻味。

进门左手处是客厅，客厅与阳台相连接，保证了充足的采光。设计时采用简约的设计风格，顶部用了造型简单的石膏线吊顶，中间安置瓷质吊灯，四周点缀以筒灯。客厅电视墙部分采用软包装饰，两侧则采用银镜，底下则安置了电视柜，通过刚直的线条和黑、白、灰三色来表现电视背景墙的丰富层次。

客厅采用乳白色、棕色相结合的沙发组合，背景墙部分则采用同种色系的壁纸，加上几幅现代风格的成品装饰画，优雅、宁静的客厅就呈现在人们面前。餐厅的一侧用不锈钢、茶镜等材质设置了装饰酒架，不仅层次感十足，又在视觉上扩大了餐厅的空间。

主卧带有阳台，采光较为充足，延续了客厅的设计风格。客厅用简单的石膏板做吊顶，原木色的饰面板装饰墙面。房间没有采用大型豪华吊灯，仅以顶部的筒灯和床头灯来调节房间的亮度。简约的现代风格的装饰画给这个空间带来了几分文艺气息。

This is an apartment with three bedrooms and two living rooms, majoring on modern concise style. It makes a lot use of log wood color veneer and soft rolls to decorate the wall to endow the space with some leisurely and easy feelings. Other than that, some parts use modern techniques as the ornaments which are full of texture, such as titanium alloy and silver mirror. The whole space is made active, bouncing and enchanting.

On the left side, upon entering the room, you can find the living room which is connected with the balcony, ensuring sufficient lighting. The design applies concise design style. The ceiling applies plaster moulding with simple format, with ceramic droplight in between and tube lamps around. Part of the TV background wall of the living room applies soft rolls as the decoration. Both sides use silver mirror. And the under part is fixed with a TV bench. The rich layers of the background wall of the TV are performed through straight line and black, white and grey colors.

The living room applies the sofa combination of milk white and brown colors. Part of the background wall applies same color wallpaper. Accompanied with several decorative paintings of modern style, elegant and tranquil living room is presented in front of our eyes. There is a decorative wine rack on one side of the dining hall made of materials such as stainless steel, tawny glass, etc. This not only enriches the sense of layers, but also visually expands the space of the dining hall.

The master bedroom had a balcony, with sufficient lighting, which continues the design style of the living room. The living room uses simple plasterboard as the ceiling and log wood color veneer to decorate the wall. The rooms do not use grand and luxurious droplights, but tube lights in the ceiling and bedside lamps to regulate the light intensity of the space. Concise decorative painting of modern style brings some artistic atmosphere.

图书在版编目(CIP)数据

摩登样板间. 第2辑. 后现代新古典/ ID Book工作室 编 —武汉：华中科技大学出版社，2013.6
ISBN 978-7-5609-8782-8

I. ①摩… II. ①I… III. ①住宅－室内装饰设计－图集　IV. ①TU241-64

中国版本图书馆CIP数据核字(2013)第056630号

摩登样板间II·后现代新古典	ID Book工作室 编

出版发行：华中科技大学出版社（中国·武汉）
地　　址：武汉市武昌珞喻路1037号（邮编：430074）
出 版 人：阮海洪

责任编辑：赵爱华	责任监印：秦　英
责任校对：王孟欣	装帧设计：吴亚兰

印　　刷：天津市光明印务有限公司
开　　本：965 mm×1270 mm　1/16
印　　张：19
字　　数：275千字
版　　次：2013年6月第1版第1次印刷
定　　价：328.00元(USD 69.99)

投稿热线：(010)64155588-8000　hzjztg@163.com
本书若有印装质量问题，请向出版社营销中心调换
全国免费服务热线：400-6679-118　竭诚为您服务
版权所有　侵权必究